PRACTICAL GUIDE TO THE MIMO RADIO CHANNEL WITH MATLAB® EXAMPLES

PRACTICAL GUIDE TO THE MIMO RADIO CHANNEL WITH MATLAB® EXAMPLES

Tim Brown
University of Surrey, UK

Elisabeth De Carvalho
Aalborg University, Denmark

Persefoni Kyritsi
Aalborg University, Denmark

A John Wiley & Sons, Ltd., Publication

Library of Congress Cataloging-in-Publication Data

Brown, Tim, 1976-
   Practical guide to the MIMO radio channel with MATLAB® examples / Tim Brown,
Elisabeth De Carvalho, Persefoni Kyritsi.
       p. cm.
   Includes bibliographical references and index.
   ISBN 978-0-470-99449-8 (cloth) – ISBN 978-1-119-94495-9 (PDF) – ISBN
978-1-119-94496-6 (oBook)
   1. MIMO systems.   I. De Carvalho, Elisabeth.   II. Kyritsi, Persefoni.   III. Title.
   TK5103.4836.B76 2012
   621.384–dc23

                                                                    2011042249

A catalogue record for this book is available from the British Library.

ISBN: 9780470994498

Set in 10.5/13pt, Times Roman by Thomson Digital, Noida, India
Printed and bound in Singapore by Markono Print Media Pte Ltd

To Rebecca

To my family

To my teachers and my students

Contents

Preface

The purpose of this book is to introduce the concept of the Multiple Input Multiple Output (MIMO) radio channel, which is an intelligent communication method based upon using multiple antennas. The book opens by explaining MIMO in layman's terms to help students and people in industry working in related areas become easily familiarised with the concept. Therefore the structure of the book will be carefully arranged to allow a user to progress steadily through the chapters and understand the fundamental and mathematical principles behind MIMO through the visual and explanatory way in which they will be written. It is the intention that several references will also be provided, leading to further reading in this highly researched technology.

MIMO emerged in the mid 1990s from work by engineers at Bell Labs as a natural evolution of long existing beamforming and diversity techniques that use multiple antennas to improve the communication link. MIMO is able to ascertain different paths over the air interface by using multiple antennas at both ends, thus creating sub-channels within one radio channel and increasing the data transmission (or capacity) of a radio link. This is a promising and important technology to meet the growing demands of high data rate wireless communications. Research has broken out worldwide, for which there is still a great need for further research and innovation in the future. Many questions have been raised over the viability of MIMO against cost and complexity though it has already been deployed in wireless local area networks and is coming back into the third generation (3G) standards for mobile cellular communications. Many people working in several related areas such as developing low cost antennas and radio transceivers need to gain understanding of and appreciation for the concept for which they will be used.

At present there is no book written in a down-to-earth manner which explains MIMO from the perspective of the antennas and the radio channel; the vast majority of books look at it from a system encoding perspective. Therefore, in the interests of those who are concerned with the way in which MIMO uses the propagation channel, this book will be valuable to any university library and also people in industry seeking to be informed about MIMO from a more visual rather than mathematical perspective. Though the reader may be new to MIMO, they would be expected to have some competence in mathematics (mainly linear algebra), foundational aspects of antennas, radio propagation and digital communications (such as modulation and coding) as well as signal processing.

Within this book is also contained, where appropriate, some MATLAB® examples that will be invaluable in helping to implement the MIMO channel models described in the book. Illustrations are also used widely within the book, especially where describing MIMO antennas and the propagation channel is concerned. It is hoped that this will give the reader an instant opportunity to begin experimenting with MIMO channel models, as well as consideration to the most appropriate channel model for their application.

We would like to acknowledge everyone who has allowed images to be used in this publication as well as the advice they have provided on certain aspects of the book.

An accompanying website containing MATLAB® code relating to this book is available at www.tim.brown76.name/MIMObook.

Tim Brown, University of Surrey, UK
Elisabeth De Carvalho, Aalborg University, Denmark
Persefoni Kyritsi, Aalborg University, Denmark

List of Abbreviations

AMC	Adaptive Modulation and Coding
AOA	Angle of Arrival
AOD	Angle of Departure
ARQ	Automatic Repeat reQuest
AS	Angular Spread
BAN	Body Area Network
BER	Bit Error Rate
BPR	Branch Power Ratio
BS	Base Station
CDMA	Code Division Multiple Access
CDF	Cumulative Distribution Function
CDL	Cluster Delay Line
COST	European Co-operation of Science and Technology
CSI	Channel State Information
CSIT	Channel State Information at Transmitter
CSIR	Channel State Information at Receiver
DG	Diversity Gain
DS	Delay Spread
DVB-T2	Second Digital Video Broadcasting Standard for Terrestrial
EDGE	Enhanced Data Rates for GSM Evolution
EPWBM	Experimental Plane Wave Based Method
FDMA	Frequency Division Multiple Access
FF	Fast Fading
GPRS	General Packet Radio Service
GSM	Global System for Mobile
HARQ	Hybrid Automatic Repeat Request
HE	Horizontal Encoding
HF	High Frequency
IEEE	Institute of Electrical and Electronic Engineers
i.i.d	Independently Identically Distributed
IQHA	Intelligent Quadrifilar Helix Antenna
ISI	Inter Stream Interference
LOS	Line of Sight
LTE	Long Term Evolution (of third generation mobile)

MAC	Media Access Control layer
MATLAB	MATrix LABoratory
MEG	Mean Effective Gain
MF	Matched Filter
MIMO	Multiple Input Multiple Output
MISO	Multiple Input Single Output
ML	Maximum Likelihood
MMSE	Minimum Mean Squared Error
MRC	Maximal Ratio Combining
MS	Mobile Station
MSE	Mean Square Error
NB	NarrowBand
NLOS	Non Line of Sight
OFDM	Orthogonal Frequency Division Multiplexing
PAN	Personal Area Network
PDF	Probability Density Function
PHY	PhYsical Layer
PIFA	Planar Inverted F Antenna
QAM	Quadrature Amplitude Modulation
QO-STBC	Quasi Orthogonal Space Time Block Coding
QPSK	Quadrature Phase Shift Keying
RMS	Root Mean Square
Rx	Receiver
SDMA	Space Division Multiple Access
SF	Shadowing Factor or Slow Fading
SIC	Successive Interference Cancelation
SIMO	Single Input Multiple Output
SISO	Single Input Single Output
SNR	Signal to Noise Ratio
STC	Space Time Code
STBC	Space Time Block Code
STTC	Space Time Trellis Code
SVD	Singular Value Decomposition
TDMA	Time Division Multiple Access
TI	Time Invariant
TRP	Total Radiated Power
Tx	Transmitter
UHF	Ultra High Frequency
UMMSE	Unbiased MMSE
UTRA	Universal Terrestrial Radio Access
V-Blast	Vertical Bell Labs Space Time
VE	Vertical Encoding
WB	Wideband
WiMAx	Worldwide Interoperability for Microwave Access

WINNER	European project Wireless World INitiative New Radio
WLAN	Wireless Local Area Network
WMF	Spatial Whitening Matched Filter
XPR	Cross Polar Ratio
ZMCCS	Zero Mean Complex Circularly Symmetric
ZF	Zero Forcing
3GPP	Third Generation Partnership Project
3G SCM	Third Generation Spatial Channel Model

List of Symbols

The following notations defined will in many cases use a subscript and/or superscript for a specific case within the book, though the physical quantity or variable that it represents remains the same.

A	Quantity of complex amplitude due to an E-field
$\mathbf{a}$	Vector with elements of complex amplitude due to an E-field
a	Quantity of complex amplitude due to an E-field
α	Angle relative to a line
B	Frequency bandwidth
$\mathbf{b}$	Bit sequence
β	Phase constant, $2\pi/\lambda$
C	Matrix for deriving phase weights
C	Capacity in bits/s/Hz
c	Speed of light, 2.98×10^8 m/s
$c()$	Codeword encryption function
D	Directivity
D_m	Maximum antenna dimension in metres
d	Separation between two points in space
$E[\mathbf{x}]$	Expected value function of vector or matrix $\mathbf{x}$
E	Quantity of complex E-field
$\mathbf{e}$	Vector with elements of complex incident E-field
F	Matrix for deriving phase weights
f	Frequency
G	Pre processing matrix
G	Antenna gain
Γ	Reflection coefficient
γ	Path loss exponent or eigen channel to noise coefficient
H	MIMO channel matrix
H	SISO channel in the case of a frequency dependent wideband channel impulse, or magnetic H-field
h	SISO channel coefficient
I	Identity matrix
K_f	Rice factor
K	Arbitrary constant

k_r	Scattering coefficient
k	Integer used for a discrete sample
L	Matrix of scattering coefficients of different polarisation
L	Loss factor
l	Spatial separation between two points in metres
λ	Free space wavelength
λ_i	Eigenvalue (used in this form where there is an integer number subscript i)
M	Integer number
m	Integer number
μ	A mean value
N	Integer number
$\mathbf{n}$	Vector containing streams of additive Gaussian white noise
n	Integer number or additive Gaussian white noise
P	Power matrix
P_{root}	Diagonal matrix of square root of power elements
P	Quantity of power
Pr {}	Probability distribution function
p	Power density at an angular point or probability quantity
p'	Normalised power density at an angular point
ϕ	Angle or phase
φ	Angle or phase
Q	Cholesky factor of covariance matrix $\mathbf{R}$
q	Complex coefficient
θ	Angle or phase
ϑ	Angle or phase
R	Correlation or covariance matrix
$\mathcal{R}$	Maximal transmission rate
R	Quantity of resistance in Ohms, covariance or transmission rate
R	Bit rate in bits per transmission
r	Physical separation between two points in metres
r_H	Channel matrix rank
ρ	Mean signal to noise ratio used to define capacity
ρ_{xy}^{Tx}	Correlation between two points x and y at a transmitter
ρ_{xy}^{Rx}	Correlation between two points x and y at a receiver
S	Matrix of singular values
S	Power density
s	Singular value
σ	Value of standard deviation
T	Time period
T_{xy}	Scattering coefficient of de-polarisation from polarisation state x to state y
t	Quantity of time in seconds
τ	Time delay
U_x	Unitary and orthogonal eigenvector matrix

$\mathbf{u}_x$	Eigenvector
u	Eigenvector element
V	Unitary and orthogonal eigenvector matrix
V	Quantity of voltage
$\mathbf{v}$	Eigenvector or vector describing voltages at an N-port network
v	Eigenvector element
v_m	Velocity of a mobile terminal
w	Digital codeword
w	Complex phase weight
$\mathbf{w}$	Vector of complex phase weights
$\mathbf{\Omega}_w$	Parameter matrix to describe the channel used in the Weichselberger model
X	Quantity of reactance in Ohms or an input data packet variable
$\mathcal{X}$	Input codeword
$\mathbf{x}$	Vector of input data streams
x	Input data stream or distance in metres
x^+	Function to return the positive or zero value of x, or zero otherwise
$\mathbf{x}^H$	Hermitian transpose of vector or matrix $\mathbf{x}$
Y	Output data packet variable
$\mathcal{Y}$	Output codeword
$\mathbf{y}$	Vector of output data streams
y	Output data stream or distance in metres
$\mathbf{Z}$	Impedance matrix
Z	Quantity of complex impedance in Ohms
$\mathcal{Z}$	Candidate codeword
$\mathbf{z}$	Candidate data stream
z	Distance in metres
ζ	Angle or phase

1

Introduction

Tim Brown and Persefoni Kyritsi

People in the modern day are on the move, either just within a home or office building or traveling from place to place. In previous years, having to go to a fixed location in order to make a telephone call was an inconvenience. Nowadays this inconvenience is minimised because the facility of a mobile phone allows someone to carry out this mundane task almost whenever or whenever they please.

As mobile communications have grown exponentially in recent years and, in parallel, the World Wide Web and its applications have spread widely, the possibility to have access to Internet, entertainment and multimedia communications wirelessly has accelerated a similar trend: people want to avoid further inconveniences of having to go to a fixed location and establish the necessary communication link to get access to the Internet. Moreover, it is desirable to avoid having to install the corresponding cables into a building or home and to avoid incurring the related costs.

Speech communication for mobile telephones was at the time a tremendous task to achieve wirelessly, and internet and multimedia applications which by nature demand a significantly higher volume of data to be transmitted both ways through the communication link pose today an equally, if not more, challenging problem.

Multiple input multiple output (MIMO) systems have emerged as an enabling technology to achieve the design goals of contemporary communication systems and has given rise to a proliferation of research activity worldwide. The technology itself has hit the public domain and it is possible for someone today to buy a state of the art wireless access point and modem with MIMO written on the box. If the wireless communications industry will be producing more and more products with this technology then there is a need for engineers to have a comprehensive guide to learning the concept. Therefore, MIMO has

Practical Guide to the MIMO Radio Channel with MATLAB® Examples, First Edition.
Tim Brown, Elisabeth De Carvalho and Persefoni Kyritsi.
© 2012 John Wiley & Sons, Ltd. Published 2012 by John Wiley & Sons, Ltd.

become a widespread research topic and is a major component on the teaching agenda in many universities delivering courses on modern-day mobile communication systems.

As an aid to the telecommunications engineer of the future, this book aims to approach the subject in a way that is both intuitive and technical. The goal is to show how MIMO systems have emerged from conventional systems, and what special technical features enable MIMO systems to transmit data through a radio environment more rapidly. To the extent possible, visual means will be used to develop understanding and intuition.

This chapter will explain where MIMO came from, what it is for and how it is used. The purpose is to help the reader establish on an abstract level the right frame of mind to read the remainder of the book. The final section of this chapter will explain the structure of the book and how it will take the reader through the different stages of learning about the subject. The same structure characterises the rest of the book: every chapter contains a summary at the end, which will provide bullet points of the fundamental facts that the reader should grasp in order to progress through the book effectively.

1.1   From SISO to MISO/ SIMO to MIMO

1.1.1   Single Input Single Output SISO

SISO stands for Single Input Single Output and has conventionally been the structure used in communications systems: in general, an 'input' is the signal transmitted from a single antenna, whereas an output is the signal received on a single antenna. Indeed, conventional wisdom tells us that cellular phones have one single antenna (e.g. visible antenna extending on one end of the phone, or patch antenna embedded on its back) and communicate with a single antenna at the base station. In any radio environment there is going to be more than one user and all the users need to have access to the cellular services at the same time. The signals to the users are then separated in time (time division multiple access, TDMA), in frequency (frequency division multiple access, FDMA), or code (code division multiple access, CDMA). In TDMA systems, all users use the same set of frequencies to communicate, but not simultaneously. For example, two users can use the same frequency but they will have allocated time slots as illustrated in Figure 1.1. In this case, the communication link switches alternately between the two users so that

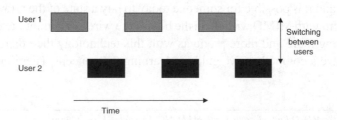

Figure 1.1   Illustration of a simple time slot allocation scheme for two users.

they can take turns to use the spectrum for equal time periods. Therefore each user's link will effectively get switched on and off for equal time periods. During the period that each user is receiving data, the data are conveyed in digital form in packets which arc then re-assembled at the receiver.

In FDMA systems, all users have access to the system at the same time, but they use different parts of the spectrum to communicate. As long as the users use different frequencies, their signals do not interfere with each other.

In CDMA systems, all users use the same spectrum at the same time, but their signals are separated in the code domain, that is, each user's signal uses a code that is specific to this user and orthogonal to other users.

1.1.2 Single Input Multiple Output, SIMO, and Multiple Input Single Output, MISO

In any of the above cases, the link between the transmit and the receive antenna is impaired by the features of the radio environment. As the user moves, the signal strength varies over the small and large scale, and at times, the quality of the link is too low to deliver data successfully. This leads to unacceptable error rates or radio link failure. In order to combat this problem, a technique known as 'diversity' has been developed. Diversity relies on the use of multiple copies of the same signal, which the receiver can combine or select from. The idea behind it is that, even if one copy of the signal is of poor quality, it is unlikely that all the copies will be so, and therefore this redundancy allows the communication quality to be maintained.

There are different flavours of diversity techniques, and they can be employed either in the downlink or in the uplink.

Depending on how the multiple copies are generated, one can distinguish different types of diversity domains. For example, one can generate multiple copies of the same signal by transmitting it multiple times, which gives rise to time diversity, or one can generate multiple copies of the same signal at different parts of the spectrum, which gives rise to frequency diversity. Moreover, one can exploit the space domain, for example when the same signal is transmitted from several base station antennas and received at a single mobile terminal (this is known as large-scale or site diversity), or a receiver has several spatially separated antennas each of which receives a different copy of the signal (this is known as small-scale diversity).

Depending on the end of the communication link which employs multiple antennas, one can distinguish between transmit diversity techniques, where multiple copies of the signal are transmitted from several antennas and their superposition is received at a single receive antenna, and receive diversity techniques, where the signal is transmitted from a single antenna and multiple copies of the signal are received at several antennas. These combinations gave rise to Multiple Input Single Output techniques (multiple transmitters, a single receiver), and Single Output Multiple Input techniques (single transmitter, multiple receivers).

Another way to classify diversity techniques is according to the way the multiple copies of the signals are exploited. In this respect, one can distinguish, in ascending order of performance, selection diversity, wherein the 'best' copy of the signal is selected; equal gain combining, wherein the multiple copies of the signal are added; and maximum ratio combining, wherein the multiple copies of the signal are weighted by appropriately selected scaling factors such that the quality of the resulting signal is optimised.

As an example, let us consider the scenario in Figure 1.4 and imagine that there is one transmitting mobile M_1 and one receiving base station that has two antennas. The signal transmitted from the mobile station is denoted as x and the signals received at the two base station antennas are indicated as y_1 and y_2. The relationship between them is

$$y_1 = h_1 x + n_1$$
$$y_2 = h_2 x + n_2$$

(1.1)

where h_1, h_2 are the channel coefficients between the mobile station and the two receive antennas respectively, and n_1, n_2 are the noise signals at the two receive antennas, which for simplicity will be assumed to be independent and of the same statistics. The base station can combine the signals from its two receive antennas to improve the quality of the signal.

Selection diversity would select the best of the two signals, that is, the one with the largest channel coefficient, and therefore the output of a selection diversity receiver would be

$$y_{sel} = \max(|h_{11}|, |h_{12}|)x_1 + n_i$$

(1.2)

where the index i is the one of the maximum channel coefficient.

Equal gain combining would simply add the two signals, after aligning their phases so that they add coherently. Therefore the base station apply the phase weights, u_1 and u_2, to output y_{equal}:

$$y_{equal} = u_1 y_1 + u_2 y_2 = (u_1 h_{11} + u_2 h_{12})x + (u_1 n_1 + u_2 n_2)$$
$$= (|h_{11}| + |h_{12}|)x + (u_1 n_1 + u_2 n_2)$$

(1.3)

If the antenna elements are within close proximity to each other, and one assumes that the magnitudes of the channel coefficients $|h_{11}|$ and $|h_{12}|$ are the same, only their phase is different, thus the equation simplifies to $y_{equal} = 2|h_1|x$. Therefore the signal received is twice that of the signal that would be received if there had been only one single antenna at the receiver. Therefore due to using an array antenna, there is a signal gain, which is known as an array gain or beamforming gain.

Maximum ratio combining would not simply add the two signals, after aligning their phases so that they add coherently, but it would scale them suitably so that stronger signals have more weight. We denote the scaling weights as u_1 and u_2 and it can be mathematically shown that, in the case of equal average noise power, the optimal weights are proportional

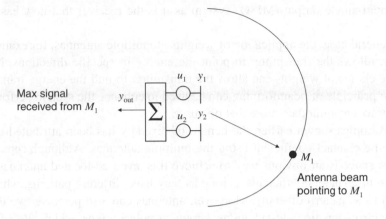

Figure 1.2 Simple illustration of beamforming gain.

to the channel coefficients $u_1 = h_1^*$ and $u_2 = h_2^*$.[1] Therefore the base station apply the scaling weights, to output y_{mrc} :

$$
\begin{aligned}
y_{mrc} = u_1 y_1 + u_2 y_2 &= (u_1 h_1 + u_2 h_{12})x + (u_1 n_1 + u_2 n_2) \\
&= (|h_{11}|^2 + |h_{12}|^2)x + (h_1^* n_1 + h_2^* n_2)
\end{aligned}
\tag{1.4}
$$

In general, the application of gains to the signals transmitted or received from multiple antennas is called beamforming. The easiest way to understand it in this context is shown in Figure 1.2, where the phase weights are optimised so as to receive the maximum signal from mobile station at a specific location. If the position of the mobile was moved to any point on the dotted semi-circle such that it was the same distance away from the array antenna, only at a different angle, the output signal received would certainly decrease because the phase factors would not correspond to the actual phases of the channel coefficients, and the signals to be added would no longer be phase-aligned. At some point, the phases might be so misaligned that the signals to be added would have opposite phases and would cancel each other. Since there is therefore only one angle at which the array antenna will receive the maximum signal from the mobile station, it is effectively forming a *beam* in a single direction, which is determined by the phase weights, u_1 and u_2.

The example shown is a case of a single input and multiple output (SIMO) system since there is one antenna at the input (i.e. the transmitter) and more than one antenna combined at the output (i.e. receiver). Given that a SIMO channel is reciprocal, we can make the array antenna the transmitter and the single mobile antenna the receiver. The beamforming gain is still valid if the phase weights are still optimised, which means we now have a

---

[1] In the case of unequal noise powers, the optimal gains are proportional to signal to noise ratio on each branch, which simplifies to the expression shown for equal noise powers.

multiple input single output (MISO) system as it is the receiver that now has only one antenna.

In the general case, the application of weights at multiple antennas, for example at the transmitter, allows the transmitter to point the energy in specific directions. Moreover, appropriate choice of weights can allow the transmitter to null the energy transmitted in others. The principle of beamforming can be used to enhance the system performance to one user, or to accommodate more that one user.

In the examples shown earlier, the benefit of diversity has been attributed to the difference in the channel coefficient using the multiple antennas. Although separating the antennas in space is an obvious way to achieve this, even co-located antennas can perceive different channel coefficients as long as they have different patterns, which would be referred to as pattern diversity. Moreover, antennas can also perceive two differently oriented electromagnetic field radiations, known as polarisations, which allow for polarisation diversity. Finally, it is possible to achieve circuit-based decoupling of antennas that are located close to each other and this is exactly the enabling factor for the use of multiple antennas on small terminals where spatial separation is not possible.

As a summary, diversity techniques have several benefits: the link quality is more stable, the average performance is improved, and we in general refer to these benefits as diversity gain. However, such techniques come at a price, namely they require more system resources. This is perhaps easier to understand on the basis of time diversity techniques, where more system time is required to transmit the same data. That extra time expenditure could have been used to transmit new data, thereby improving the system efficiency. The cost associated with the use of multiple antennas relates to space considerations (a terminal should be large enough), hardware concerns (a more complicated receiver or transmitter is required) and price issues (the additional complexity increases the device price). Another disadvantage is that diversity is a process of diminishing returns. This means that the benefit of adding for example a third antenna over having only two antennas is smaller than the benefit from going to two antennas from a single antenna. Moreover, the condition for diversity techniques to be effective is that the copies of the signal have to be independent, so as to minimise the probability that they all face simultaneously bad propagation conditions.

1.1.3 *Multiple Input Multiple Output, MIMO*

Diversity, and specifically space diversity, relies on the use of multiple antennas at one end of the communication link, either at the transmitter or at the receiver side. The evolution of the diversity ideas has been the use of multiple antennas at both ends of the communications link.

The additional benefit that can be drawn from the use of multiple antennas on both sides of the communications link is so-called spatial multiplexing, that is, the ability to send several data streams simultaneously. Indeed in the examples shown in the previous section, the same data stream exploits the benefit of diversity. Let us take the example

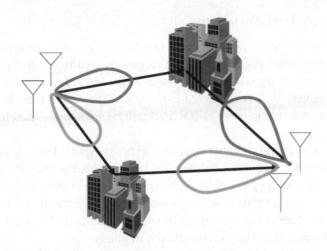

Figure 1.3   Illustration of the principle of a simple 2x2 MIMO link.

of a system with two transmitters and two receivers, each transmitter sending a separate data stream. As shown before, we can perform beamforming to each of the transmitters. If we beamform to both transmitters simultaneously, and the two beams are sufficiently separated, it would be possible to obtain the two streams without them interfering with each other. Indeed, for MIMO to work, it has to be possible for the beams formed at both ends to be able to distinguish different angles from which the antenna can transmit or receive. A simple example is shown in Figure 1.3 where effectively two paths have been created due to reflections of buildings at two different locations and the beams have been formed to transmit and receive down the two orthogonal paths.

Even though MIMO has been in the public domain for several years, it has only recently come into the production line for wireless communication devices. There are two main reasons why this has taken so long. First of all the practical aspects of implementing MIMO have been difficult in terms of building suitable antennas and also the implementation of the complex radio device has eased. Secondly, despite the fact that the cost of such a technology is still not economic, users are now prepared to pay the price if it delivers the desired services over the restricted bandwidth and is a cost-effective option.

1.2   What Do We Need MIMO For?

There are two perspectives from which we can answer this question. The first is from the perspective of a single user, wishing to increase the data rate of the wireless link between their mobile device and an access point or base station. The second is from the perspective that there are normally several users in a wireless system, communicating wirelessly at the same time and using the same frequency. We will therefore address these two points separately.

1.2.1   The Single User Perspective

The most obvious way to increase the rate at which data can be transmitted would be to increase the transmitted power so that the received signal is much stronger. However, transmitting high power signals from a mobile device has safety issues, as well as battery issues: the battery will be depleted extremely fast and the time between charges (which is a common factor in user satisfaction) will be shortened. Therefore this is not a good option.

Another way of enabling a higher data rate to be transmitted through a wireless link is to increase the radio frequency bandwidth. However, at the frequencies suitable for communication links, namely the parts of the spectrum where there are no other conflicting uses and where the propagation losses are not prohibitively high, the bandwidth is extremely precious. There are major financial and legal obstacles involved and it is simply not possible to increase the allocated bandwidth infinitely.

It turns out that MIMO systems have the ability to increase the data rate of the transmitted data without increasing the transmit power and by using the same amount of frequency resources.

1.2.2   The Multiple User Perspective

In any radio environment there is going to be more than one user and all of them need to support simultaneous links. As indicated earlier, the users have conventionally been separated in the time, frequency or code dimensions in order to avoid interference among them.

We therefore need to find methods of accommodating multiple users simultaneously within the constraints of limited frequency bandwidth, and multiple antennas come to the rescue. As explained earlier, MISO and SIMO techniques allow the formation of beams that point the energy in specific directions. By forming two beams pointing in two directions, one can effectively send different signals to two users.

As an example, let us consider one of the simplest forms of antenna array at a receiving base station, while there are two mobile devices at different locations, each transmitting a signal at the same frequency to the receiver illustrated in Figure 1.4. The mobile users, M_1

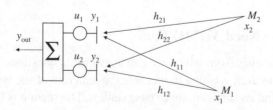

Figure 1.4   Illustration of a basic two element array antenna used to mitigate co-channel interference.

and M_2, are simultaneously transmitting signals x_1 and x_2 respectively. The superposition of the two signals at each of the two receiving the two antennas, y_1 and y_2, can be derived by the following equations:

$$y_1 = h_{11}x_1 + h_{21}x_2$$
$$y_2 = h_{12}x_1 + h_{22}x_2$$

(1.5)

In reality, there is also receiver noise added to the signals y_1 and y_2, which has for simplicity been omitted.

In the above equation, h_{11} is the complex channel coefficient between mobile M_1 and the receiving antenna 1. Likewise h_{21} is the complex channel coefficient between mobile M_2 and the receiving antenna 1. This works the same way for mobile M_2 with channel coefficients h_{21} and h_{22}.

The receiver can apply weights u_1 and u_2 on the two received signals, y_1 and y_2 and then combine them before reaching the output as shown in Figure 1.4. The resulting signal, y_{out} would be:

$$y_{\text{out}} = u_1 y_1 + u_2 y_2 = (u_1 h_{11} + u_2 h_{12})x_1 + (u_1 h_{21} + u_2 h_{22})x_2$$

(1.6)

The weights can be set appropriately so that the signal contains only terms with x_1 and not x_2, which means only the signal from mobile M_1 is received, while the signal from M_2 is suppressed. For this to be possible, the following criteria therefore has to be met (assuming the channel coefficients are appropriately normalised):

$$u_1 h_{11} + u_2 h_{12} = 1$$
$$u_1 h_{21} + u_2 h_{22} = 0$$

(1.7)

This a simple system of linear equations which can easily be solved to derive the weights u_1 and u_2 that will help isolate signal x_1.

A further step to consider applying also a second set of weights that can help isolate signal x_2 in a simlar fashion as shown in Figure 1.5. By the application of two sets of weights, the receiver has essentially formed two beams, such that y_{out1} only receives from M_1 and y_{out2} only receives from M_2. For simplicity, this technique is referred to as Space Division Multiple Access (SDMA). Therefore MIMO can be seen as an evolution of MISO and SIMO that includes the ability to handle multiple users as well as providing a higher data rate communication link.

Although the selection of the weights shown above to eliminate the interference from the other data streams is not the only, and possibly not the best, way to isolate the data transmitted by the two users, the example is a good illustration of the potential that MIMO systems have to accommodate more than one user simultaneously.

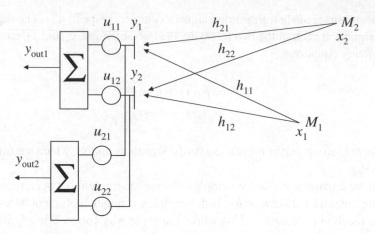

Figure 1.5 Illustration of a basic two element array antenna used for space division multiple access.

1.3 How Does MIMO Work? Two Analogies

Before considering how a MIMO radio channel works using antennas intelligently, we will take two analogies that help to understand the concept in a more visual way. These two ways are going to be analogous to the single and multiple user perspectives presented in the previous section.

1.3.1 The Single User Perspective

Let us first consider the *football match problem*. Supporters on their way to a football match arrive on a bus, train or tram where they exit the station. In the ideal case, they take a wide, direct path to move directly over to the stadium entrance in an orderly queue, thus all walking there as quickly as possible as illustrated in Figure 1.6.

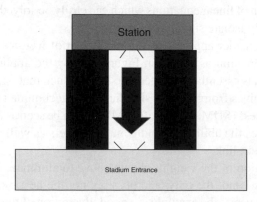

Figure 1.6 Illustration of a perfect scenario of supporters entering a football stadium.

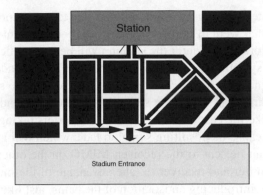

Figure 1.7   Illustration of a real scenario of football supporters entering a stadium without crowd control.

In practice, however, such wide direct maths between the station and the stadium might not exist. Instead, the station may be some distance away and supporters can take several possible routes down different streets to arrive at the same destination as the illustration in Figure 1.7 shows. The picture shows that there are several points where the separate paths that supporters take begin cross, and the traffic can collide. This will slow down the rate at which supporters will enter the building and some will inevitably be stuck for quite some time.

So that a crowd of supporters can walk to and from a station to a stadium quickly and efficiently, it is necessary to enforce crowd control, which will determine which routes the supporters take and which entrances or exits they use. One such example of crowd control is shown in Figure 1.8, where two entrances are opened at the stadium and two exits are opened at the station. Furthermore barriers have been implemented using the dotted lines shown so that the paths between the station exits and the stadium entrances do not collide.

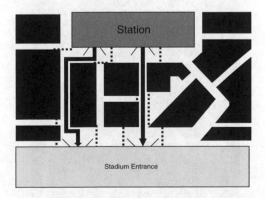

Figure 1.8   Illustration of a real scenario of football supporters entering a stadium with crowd control.

Even though each route individually might only be able to accommodate a portion of the traffic, the collisions that were in Figure 1.7 have now been prevented, and cumulatively the rate at which supporters will enter the stadium will be more efficient.

For wireless radio systems, the transmitted radio signal will encounter many objects that will cause the signal to reflect, refract and diffract causing the signal to arrive at the receiver from many directions. Thus the signal has taken many paths and this is known as *multipath*, which like the case of many football supporters taking several paths to the same destination will cause the resulting signal to fade, which will at the same time limit the rate of data that can be sent to the receiver. MIMO on the other hand uses multiple antennas at the transmitter and receiver to take advantage of the multipath and transmit more data by arterially introducing an isolation of the paths, just like in the case of crowd control at a football match.

1.3.2   The Multiple User Perspective

We will now consider the multiple user perspective by taking the example of the *cocktail party effect* shown in Figure 1.9. There are three groups of people in the picture amongst several other groups in a large hall and everyone is enjoying themselves talking loudly and laughing. Anyone who has lived through such a situation can attest that, despite the ambient noise, one can concentrate on the conversation within one's group. With a bit of effort, one can isolate what the other person is saying, and sometimes, one can even do so for people that are farther away and whose voice is almost lost in the noise. This is because, although the human ear is capable of listening to everything around it no matter what direction the sound is coming from, it can also concentrate on a specific source

Figure 1.9   A cocktail party example of the multiple user perspective.

in a manner very similar to a camera focussing on a single point. In a sense our ears are an array of two elements that can concentrate on the person we are talking to and suppress the noise from others conversing near to us. Therefore the group in the middle of Figure 1.9 can communicate with each other while suppressing the acoustic interference of the two groups either side.

In MIMO, arrays of antennas rather than arrays of ears are used to suppress radio interference at the same frequency. The antenna arrays at both ends of a MIMO link will also concentrate radio signals into different directions to concentrate data into different paths and allow more data to be transmitted.

1.4   Conditions for MIMO to Work

As shown earlier, the derivation of suitable weights that can help isolate the data streams (either from multiple users or from a single user equipped with multiple antennas) relies on the solution of a system of linear equations that involve the channel coefficients. Basic algebra indicates that there cases where the solution is feasible and others where it is not. These conditions translate into requirements for the channel coefficients

To analyse more closely how the channel coefficients should be in order for MIMO to work, the three scenarios can be considered in Figure 1.10. First of all in Figure 1.10 (a), there are are no objects in the vicinity of the transmitting antennas on the left and receiving

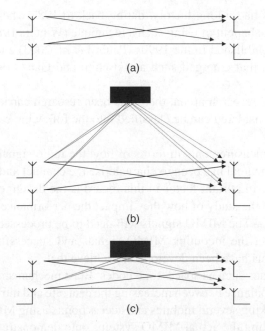

Figure 1.10   Illustration of the scattering scenarios with which MIMO will operate.

antennas on the right, just free space. Therefore if the transmit antennas are close together, and the receive antennas are also close together, then it is difficult to make two beams to isolate the two signals. In this case MIMO communication would not be possible.

If an object is introduced to create some reflected paths as illustrated in Figure 1.10 (b), different paths can be created. The object is known as a *scatterer*, and beams can be formed so that one channel is created that exploits the reflected paths, while the other channel will use the direct link from the transmitter to the receiver. It can now be seen how MIMO will exploit the multipath to create more channels through which it is able to transmit data more rapidly. If there were more scatterers present and more antennas at each end, this would mean there is more multipath available to exploit and also more separate channels could be created to enhance the separation of the signals.

A final point to note regarding the exploitation of multipath by MIMO is that it must also meet certain conditions. Taking Figure 1.10 (c), for example: the scatterer is now very close to the direct path between the transmitter and receiver. Because the paths are then much closer and therefore much more similar to each other (i.e. they are highly correlated) then it becomes too difficult to form unique beams at both ends from which orthogonal channels can be created. Therefore the position of the scatterers as well as their density are very important. This is often quantified by what is known as *scattering richness*, which therefore will be of great benefit to MIMO.

1.5   How Long Has MIMO Been Around?

Although MIMO has had a long history, dating back to 1987 when the first paper was published showing the benefit of using such a technique (Winters 1987). Some of the first related patents were published in the 1990s (Paulraj *et al.* 2003) and then later on some widely cited journals that emerged such as Foschini and Gans (1998) and Telatar and Tse (2000).

In the international research arena, there has been research carried out in MIMO and the research topics considered can be classified into the following categories:

- **Radio** – Channel measurements in terms of how the radio signal propagates through the environment. the goal has been to characterise the channel and derive models of its behaviour. Another important factor in this regard is the design of the array antennas used in MIMO and the study of how they impact the performance.
- **Signal processing** – The MIMO signals will need to be processed such that they have algorithms to detect the incoming MIMO signals and successfully decode them. A variety of algorithms have been developed to achieve that.
- **Coding** – Techniques such as space–time codes, Blast mechanisms and eigen tracking aim at achieving a balance between increasing the data rate and introducing redundancy.
- **Networking** – Linking several mobiles and access points using MIMO simultaneously.
- **Hardware** – Building the actual MIMO systems onto demonstrators, radio frequency (RF) chips and digital signal processing (DSP) chips etc.

The main focus of this book will be the radio and the signal processing involved in MIMO digital communications, including the theoretical limits imposed by information theory.

1.6   Where is MIMO Being Used?

Currently, MIMO is being used in some of the latest wireless routers to enhance the capacity of the wireless local area network (WLAN). This is shown in Figure 1.11 with the multiple antennas spatially separated. Many wireless routers such as this are normally defined under the Institute of Electrical and Electronic Engineers (IEEE) 802.11 standards (IEEE802.11). In the future MIMO will be used in cellular standards such as IEEE 802.16 (known as WiMAX) and Third Generation Partnership Long Term Evolution (3GPP LTE) (3GPP), which will be used in mobile communication devices that have MIMO capabilities. Such a technology will be essential in delivering the wireless broadband and multimedia services that these standards demand.

Figure 1.11   Illustration of a wireless router using MIMO. Reproduced by permission of Belkin Ltd.

1.7   Purpose of the Book

This book has opened up with a brief introduction to MIMO, stating both its function and its purpose. The theory behind MIMO will be explained in greater mathematical detail in the next chapter, paying particular attention to the information theory capacity limits. The book will then consider algorithms with which one can actually allow MIMO to work in practice and measure its performance in the third chapter. As MIMO performance depends on the propagation conditions, Chapters 4 and 5 will respectively address the radio channel and the design of the suitable antennas. Finally the book will conclude by giving specific examples of how MIMO is being used today, future trends and further research still to be carried out.

2

Capacity of MIMO Channels

Elisabeth De Carvalho

This chapter is dedicated to the capacity of MIMO channels. Capacity is a performance measure for digital communication systems. It is the maximal transmission rate for which a reliable communication can be achieved. If the transmission rate gets larger than the capacity, the system 'breaks down' and the receiver makes decoding errors with a non negligible probability. Capacity is the primary tool to characterise the performance of MIMO systems and it also serves in practical system as a guide to properly design the transmitted signals as well as the processing of the received signals.

Wireless communications exhibit different characteristics and performance according to the propagation environment. Those characteristics have to be carefully taken into account when defining capacity. In this chapter, capacity is described according to two factors impacting the performance:

1. *The knowledge of the channel at the transmitter and receiver.* A common assumption, adopted throughout the chapter, is that the channel is perfectly known at the receiver. At the transmitter, two different types of channel knowledge are considered: either the instantaneous value of the channel is known or only its distribution is known.
2. *The nature of the wireless channel.* We treat three kinds of channels: the time invariant channel, the fast fading channel (the codewords spread over many channel variations) and the slow fading channel (the channel is constant across a given codeword).

We present capacity results for the following three scenarios which are the most common and practical ones.

Practical Guide to the MIMO Radio Channel with MATLAB® Examples, First Edition.
Tim Brown, Elisabeth De Carvalho and Persefoni Kyritsi.
© 2012 John Wiley & Sons, Ltd. Published 2012 by John Wiley & Sons, Ltd.

	Distribution of CSIT Instantaneous CSIR	Instantaneous CSIT Instantaneous CSIR
Time Invariant Channel		Case I (section 2.5)
Fast Fading Channel	Case II (section 2.6)	
Slow Fading Channel	Case III (section 2.7)	

The channel state information at the transmitter (CSIT) is the information about the channel available at the transmitter while the channel state information at the receiver (CSIR) is the information about the channel available at the receiver. For each scenario, we build on the single input single output (SISO) and single input multiple output (SIMO) or multiple input single output (MISO) channels to present the multiple input multiple output (MIMO) channel. In particular we highlight the performance boost brought on by the multiplexing capabilities of systems with multiple antennas at both the transmitter and the receiver.

2.1   Some Background on Digital Communication Systems

Digital communications consist of the transfer of *bits* (0 or 1) from a transmitter to a receiver. The bits carry the information to be communicated, so the first important question is how the information is converted into bits and how bits can be transported through the communication medium.

2.1.1   Generation of Digital Signals

Some signals are in digital form: files in a computer or digital photographs. No further processing is needed in general before converting the signals for transmission. However, most signals are analogue, like voice or sound.

The conversion of analogue signals into digital signals is based on sampling and quantisation. The analogue signal is first sampled. The sampling rate should be sufficiently high, so that the sampling does not entail any loss of information, meaning that the original continuous time signal can be recovered from the sampled signal (Nyquist theorem). In the quantisation step, the samples are quantised to a finite number of levels. The sampled signal takes only a finite number of values. The set of possible values forms a *finite alphabet*. The number of elements in the finite alphabet is generally a power of 2 and hence can easily be represented by a sequence of bits.

To reduce the amount of information to be transmitted, the sampled data goes through a *source coding* or *data compression* step. The purpose is to remove the redundancy in the data and to minimise the number of bits used to represent the data without changing the perceived quality of the initial signal.

2.1.2 Conversion/Formatting for Transmission

The purpose of this conversion is to adapt the signal to the communication medium by: (1) providing protection against channel impairments, (2) converting the digital signal into a signal that can be physically sent through the medium.

When travelling from the transmitter to the receiver, the signals are affected by various impairments (e.g. thermal noise at the receiver, channel fades in wireless communications). To provide protection against those communication impairments, redundancy is introduced in the data by means of a *code*. This process is called *channel coding*. The code maps the data into a new finite alphabet: the input of the code consists of a set of bits and its output is a larger size set of redundant bits. Each possible output of the code is a *codeword*. The idea is that if a part of a codeword is affected by the channel impairments, then the redundant part might not be.

Next, the coded signal is converted to be physically transported through the medium: it is converted back to an analogue form. This conversion operation is called the *modulation*. The modulation uses its own finite alphabet: for example, in a 4-QAM (or QPSK) constellation, the finite alphabet comprises 4 elements. Those elements are called *symbols*. In some advanced schemes, channel coding and modulation are performed jointly.

2.1.3 Complex Baseband Representation

In wireless communications, during the modulation process, the signal is shifted to some carrier frequency to better fit the medium properties. The spectrum of the modulated signal is centered around the carrier frequency, for example, 2 GHz for current wireless systems. The modulated signal is a bandpass signal. In the demodulation process, the signal is brought back to low frequencies around 0 (baseband signal) and sampled. For a linear modulation, this baseband signal is conveniently represented using an equivalent complex baseband signal. This representation is used throughout Chapters 2 and 3. For a simple SISO narrowband channel model, the signal can be simply written as:

$$y(k) = h(k)x(k) + n(k) \qquad (2.1)$$

All the quantities are complex: $x(k)$ is the symbol transmitted at sampling time kT (T is the sampling period). $h(k)$ is the channel and $n(k)$ is the noise at the receiver.

2.1.4 Decoder

Assuming that the channel is known at the receiver, the optimal decoder is based on the maximum likelihood (ML) criterion. ML lays in finding the codeword that fits the received signal the best. Based on the complex baseband representation of the signal, the

ML criterion for a SISO narrowband channel is:

$$\min_{\{c(1),\ c(2),...,\ c(N)\}} \sum_{k=1}^{N} |y(k) - h(k)c(k)|^2$$

The ML cost function should be evaluated at each possible codeword of the alphabet denoted as $\{c(1), c(2), \dots, c(N)\}$. A successful decoding is based on the ability of the receiver to distinguish the transmitted codewords from each other and hence on the minimal distance between codewords. Continuing with the SISO example, if the additive noise introduces too large a perturbation, then one codeword could be mistaken for another and decoding errors occur. The minimal distance between codewords depends on the transmit power, the bit rate and the length of the codeword. For a fixed transmit power and codeword length, increasing the bit rate means decreasing the distance between symbols (it is similar to squeezing more points in a confined space). In such a case, the error probability increases.

2.2   Notion of Capacity

2.2.1   Abstract Communication System

To introduce the notion of capacity, we consider a simple abstract communication system, where a sequence of bits $\mathbf{b} = \{b(1), b(2), \dots, b(N_b)\}$ is communicated to a receiver. The case of a SISO channel is depicted in Figure 2.1. At the transmitter side of the SISO system, an encoder maps the bits into a codeword $\{x(1), x(2), \dots, x(N)\}$. The symbols $x(k)$ are selected from a finite alphabet set and are sent one after another through the wireless channel. A codeword contains N symbols. At the receiver, a decoder demaps the received codeword $\{y(1), y(2), \dots, y(N)\}$ to recover the transmitted message. The decoded bit sequence is denoted as $\hat{\mathbf{b}}$.

In the MIMO case (Figure 2.2), a space dimension is added due to the presence of multiple antennas. The encoding is done across time, like for SISO, but additionally across space. Such a codeword spanning two dimensions can be represented as a matrix

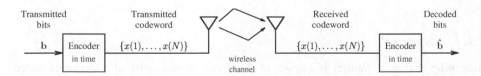

Figure 2.1   General communication system: SISO case.

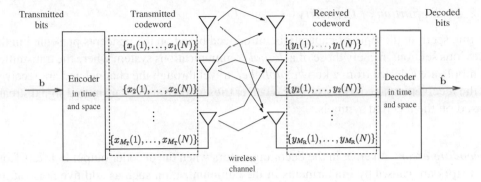

Figure 2.2 General communication system: MIMO case.

as follows:

$$\underset{\text{space}}{\Big\downarrow} \quad \begin{bmatrix} x_1(1) & x_1(2) & \cdots & x_1(N) \\ x_2(1) & x_2(2) & \cdots & x_2(N) \\ \vdots & \vdots & \vdots & \vdots \\ x_{M_T}(1) & x_{M_T}(2) & \cdots & x_{M_T}(N) \end{bmatrix} \qquad (2.2)$$

$\overset{\text{time}}{\longrightarrow}$

Line i in this matrix, $\{x_i(1), x_i(2), \ldots, x_i(N)\}$, represents the symbols sent from antenna i at time instant $1, \ldots, N$. Column j, $\{x_1(j), x_2(j) \ldots x_{M_R}(j)\}$, represents the symbols sent at time j from all antennas.

To the transmitted codeword corresponds a received codeword spanning over time and space. It is represented in matrix form as:

$$\begin{bmatrix} y_1(1) & y_1(2) & \cdots & y_1(N) \\ y_2(1) & y_2(2) & \cdots & y_2(N) \\ \vdots & \vdots & \vdots & \vdots \\ y_{M_R}(1) & y_{M_R}(2) & \cdots & y_{M_R}(N) \end{bmatrix} \qquad (2.3)$$

Line i in the matrix, $\{y_i(1), y_i(2), \ldots, y_i(N)\}$, represents the symbols received at antenna i at time instant $1, \ldots, N$. Column j, $\{y_1(j), y_2(j) \ldots y_{M_R}(j)\}$, represents the symbols received at time j at all antennas.

The *system bit rate* is the transmission rate before encoding, that is, the number of bits in the message **b** transmitted per time unit. Consider the example of a simple repetition code replicating each bit of **b** twice. If the bit rate after the encoder is R bits per transmission, then the system bit rate is R/2 bits per transmission.

2.2.2   Definition of Capacity

In this section, the topic of capacity is introduced based on the concepts presented in the previous sections, namely those of a *digital* communications system where the transmitted symbols are selected from a known *alphabet*, travel through the channel and are received at the receiver impaired by *noise*. The receiver tries to recover the transmitted signal stream based on the received symbols.

Decoding error: An error in the decoding occurs when the decoder output $\hat{b}$ differs from **b**. Errors are caused by impairments in the communication such as additive noise at the receiver or an unexpected deep fade for a wireless channel. As the channel impairments are random, so is the event of a decoding error. Hence, we define the probability of error, where the probability is taken over the channel impairments. The error probability cannot be exactly equal to zero in general. Indeed, the channel impairments cannot be exactly predicted and can be arbitrarily large; for example, at a given time, the additive noise at the receiver can be very large, thus making the receiver unable to decode the data impaired by this large noise sample.

Error probability brought arbitrarily closed to zero: If the error probability cannot be made equal to zero, it can however be brought arbitrarily close to zero by adjusting certain parameters. Such a parameter is the transmit power, where provided that the power in the impairments remain the same, the error probability gets decreased by increasing the transmit power. An asymptotically large transmit power results in an error probability asymptotically close to zero.

Due to system constraints or regulations, the transmit power is limited in general. Therefore, the error probability cannot be brought down close to zero by increasing the transmit power. However, if the transmission rate is smaller than a certain limit, known as the system capacity, it is still possible to bring the error probability arbitrarily close to zero, even if the transmit power is limited. When coding over a very large block of data where many realisations of the impairments occur, the channel impairments become averaged out, thus decreasing their randomness or unpredictability. Because the impairments become quasi-deterministic, a code can be constructed resulting in a very low error probability.

However, when the bit rate is higher than the system capacity, no code can be constructed that can compensate for the channel impairments. Even if the impairments are averaged out, they are still detrimental to the communication. When the bit rate exceeds the system capacity, the distance between codewords becomes too small for the received codewords to be distinguishable. Whatever the efficiency or length of the code, the error probability cannot be brought down close to zero.

Next, we give a summary of the main points from the previous discussion:

Capacity definition and key properties.

Definition: The capacity of a communication system is the maximum transmission rate for which the decoding error probability can be brought arbitrarily close to zero.

Capacity is also defined based on the notion of reliability. A communication at rate R is said to be *reliable* if one can design a code at rate R that makes the error probability arbitrarily small. Capacity can be defined as the maximum transmission rate for which a reliable communication can be achieved.

Key properties:

▶ A reliable communication can be achieved for any transmission rate smaller than or equal to the capacity.

▶ Conversely, if the transmission rate is strictly larger than the capacity, then there exists no code which brings the error probability arbitrarily close to zero (the error probability is lower bounded by a nonzero value).

▶ Capacity is achieved by coding over an arbitrarily long block of data. Furthermore, no constraint is assumed on the complexity of the coder or the decoder. But, in reality, such codes require complex coding/decoding procedures. At last, in practical digital communication systems, the input symbols belong to a constellation with a finite number of possible values. In the information theory framework, the input symbols belong to a continuous distribution. Thus, capacity achieving codes cannot be implemented in practical systems in general. The capacity is in fact a performance limit for reliable communications which is achieved under asymptotic conditions.

▶ Capacity is given under a transmit power constraint. Indeed, if the transmit power is not limited, and no matter what the transmit rate is, the error probability can always be made small by increasing the transmit power: this means that the capacity is infinite.

2.2.3 Capacity Achieving Transceivers

Although we have defined the channel capacity as an upper limit to the system performance, the most interesting question for the system designer is how this limit can be achieved. Therefore, the study on capacity includes not only the maximal transmission rate for reliable communications, but also an optimal transmitter and receiver structure able to achieve capacity. For an optimal reliable MIMO communication, the following transmission parameters should be set across time but also across space:

• *Power allocation in time and across space*: The transmit power is constrained in space such that usually the sum of transmit powers from each transmit antenna is limited to a maximal value. The transmit power is also constrained in time such that the transmit power averaged over time is also limited to a maximal value. To maximise performance, the power has to be properly jointly distributed in space and in time.

- *Transmission rate across time and space*: Capacity gives the sum of the rates from all transmit antennas. When an optimal transmitter structure is known, it is also possible to determine individual rates across space as well as their evolution in time.
- *Coding across time and space*: Depending on the channel characteristics, the optimal coding is done across many channel fades or only across one channel fade. Furthermore, coding might also be done across space.
- *Data correlation across time and space*: The symbols composing the codes achieving capacity might exhibit correlations in time and also in space. This correlation matches the channel: it matches either the realisation of the channel when it is known at the transmitter or the channel distribution.

The optimal decoding at the receiver is the Maximum Likelihood (ML) decoder, which can be computationally complex, especially when it involves a joint decoding of signals from multiple antennas. In some cases, the decoding can be simplified. For time invariant MIMO channels, we describe a capacity achieving scheme where ML decoding per codeword is optimal. For fading channels, a joint processing between streams is necessary; however, ML decoding can be replaced by a less complex decoding technique based on successive interference cancellation.

2.3   Channel State Information and Fading

Capacity was first derived by Shannon for *time-invariant* SISO channels where the channel is constant in time and the transmitter knows its value. For wireless channels, the hypotheses are different. First of all, the channel varies in time and frequency. Secondly, the channel variations imply a different level of knowledge about the channel at the transmitter and receiver. For example, if the channel varies very quickly, it is not always possible to acquire a good estimate of the channel. Capacity varies according to the nature of the wireless channel and the channel knowledge.

- A wireless channel varying in time is said to be *fading*, and two distinct cases can be distinguished depending on the duration of the codeword relative to the channel variations: the fast fading and the slow fading channel.
- The information available about the channel is called Channel State information (CSI) and can take different forms: for example, the channel coefficients, the modulus of the channel, the statistics of the channel or the noise variance at the receiver.

2.3.1   Fast and Slow Fading

The channel variations or fades are due to movements of the transmitter or receiver as well as movements of objects along the propagation paths (see Chapter 4). The performance

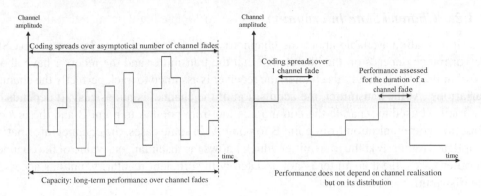

Figure 2.3   Behaviour of a fast fading and slow fading channel (single link).

of a wireless system depends on the type of fading affecting the communication. In some cases, the channel varies very fast so that the coding horizon spans many channel fades. In other cases, the channel varies very slowly and coding spans over a single channel fade. The coherence time is a useful measure to characterise the behaviour of fading channels. In short, the coherence time is the time duration before a channel varies significantly.

For the purposes of the capacity analysis, within the coherence time, the channel is considered as constant. In a point to point MIMO communication, the channels between each antenna pair, or subchannels, have the same statistics. So the coherence time of each subchannel is also the coherence time of the MIMO channel. The distinction between fast fading and slow fading is based on the channel coherence time, as described below and in Figure 2.3 for the SISO case.

Fast fading channel. In a fast fading channel, the coding delay is much larger than the coherence time. Capacity characterises performance over a time period that includes an infinite number of fading states and represents a long-term average over the channel fades. Capacity achieving coding schemes are spread over a large number of fading states in general.

Slow fading channel. In a slow fading channel, the coding delay is much smaller than the coherence time. Therefore, the channel is assumed to remain constant over the duration of a codeword.

It is important to understand the difference between a slow fading channel and a time invariant channel for which the notion of capacity was originally explored. In the latter case, the channel is always constant and capacity is given for this constant channel value. In the slow fading case, the channel can vary randomly. Capacity characterises the optimal performance of a coding scheme experiencing only one channel fade but where the channel is a random variable. Capacity of a slow fading channel does not depend on the value of channel fades but on the distribution of the fading.

2.3.2   Channel State Information

The information available about the channel is called Channel State Information (CSI). Performance depends on the knowledge that the transmitter and the receiver have about the channel. Throughout this chapter, the receiver is assumed to track perfectly the channel variations. At the transmitter, the acquisition of the channel is not as easy. It depends on the method used to separate the communication from point A to point B and the inverse direction communication from point B to point A. In some cases, the channel information is updated regularly at the transmitter which knows the instantaneous value of the channel. In other cases, the transmitter is not updated with sufficient regularity, but has access to the distribution of the channel.

CSI at the receiver (CSIR). The CSI at the receiver includes all the channel coefficients of the links going from the transmitter to the receiver. This information is necessary so that the optimal ML decoding can be performed. In some case, a simpler but still optimal receiver such as the minimum mean square error (MMSE) receiver can be used requiring additionally the knowledge of the noise variance. The receiver is assumed to perfectly track the CSI. In practical systems, the channel and noise variance are typically estimated by using pilot symbols embedded in the signal sent from the transmitter.

CSI at transmitter (CSIT). The CSI at the transmitter is more difficult to obtain in general. Its acquisition depends on the duplexing mode, that is, the method used to separate the communication from point A to point B and the inverse direction from B to A. In time division duplex (TDD), the direct and inverse communications use different time slots while in frequency division duplex (FDD) they use different frequency bands.

In wireless systems, the mechanisms for CSIT acquisition rely mainly on feedback or channel reciprocity. Let us call A the transmitter and B the receiver. The goal is for A to get an estimate of the channel going from A to B. The following two methods could be used to achieve this:

- *Feedback*: The channel is estimated at B using pilot symbols embedded in the signal sent from A to B. The CSI is then fed back from B to A.
- *Channel Reciprocity*: The channel reciprocity principle states that the channel from point A to point B is identical to the channel from B to A if the channel is measured at the same time and same frequency.

A feedback reporting mechanism can be used for both TDD and FDD. For FDD, only feedback can be used and not channel reciprocity as the direct link and inverse link do not use the same frequencies. Channel reciprocity can be used in TDD. The channel from A to B can be estimated at A using pilot symbols embedded in the signal sent from B. Using the reciprocity principle, this estimate is also an estimate for the channel from A to B. In some systems, only B is able to estimate the channel from A to B: for example, due to system constraints, B cannot send pilots but A can. Then, the estimate of the channel from A to B is fed back from B.

In this chapter, capacity is given for the following two cases.

1. The receiver knows the instantaneous value of the CSIR. The transmitter knows the instantaneous value of the CSIT.
2. The receiver knows the instantaneous value of the CSIR. The transmitter does not know the instantaneous value of the CSIT but knows its distribution.

The coding scheme can be adjusted according to the available knowledge at the transmitter about the channel to achieve a reliable communication.

2.4 Narrowband MIMO Model

Although an exact definition of a narrowband channel will be provided in Chapter 4, for the purposes of this chapter it suffices to define a narrowband channel a channel where the channel coefficient between each transmitter and each receiver is a complex scalar. A multiple antenna system with M_T transmit antennas and M_R receive antennas is depicted in Figure 2.4. The channel between each transmit and receive antenna pair is assumed to be narrowband. The channel coefficient at time k between transmit antenna i and receive antenna j is denoted $h_{ji}(k)$.

At time k, the signal $x_i(k)$ is transmitted from antenna i. At receive antenna j, the received signal is denoted as $y_j(k)$ and the additive noise as $n_j(k)$. $x_i(k)$, $y_j(k)$ and $n_j(k)$ are all assumed to be complex quantities. M_T signals are transmitted forming the vector input signal $\mathbf{x}(k)$ and M_R signals are received forming the vector output signal $\mathbf{y}(k)$. The

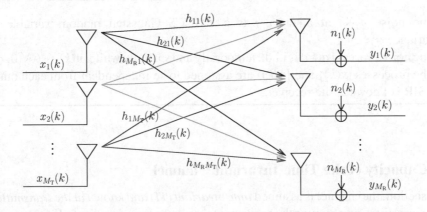

Figure 2.4   MIMO model with M_T transmit antennas and M_R receive antennas.

M_R additive noise signals are grouped in the vector $\mathbf{n}(k)$. Those quantities are defined as:

$$
\mathbf{x}(k) = \begin{bmatrix} x_1(k) \\ \vdots \\ x_{M_T}(k) \end{bmatrix} \quad
\mathbf{y}(k) = \begin{bmatrix} y_1(k) \\ \vdots \\ y_{M_R}(k) \end{bmatrix} \quad
\mathbf{n}(k) = \begin{bmatrix} n_1(k) \\ \vdots \\ n_{M_R}(k) \end{bmatrix}
\tag{2.4}
$$

The notations $\{y(k), k \in \mathbb{N}\}$, $\{x(k), k \in \mathbb{N}\}$ and $\{n(k), k \in \mathbb{N}\}$ denote random processes, that is, a collection of random variables in time.

The noise at antenna j, $n_j(k) = \mathrm{Re}(n_j(k)) + \mathrm{jIm}(n_j(k))$, is a complex zero mean Gaussian random variable with variance σ_n^2. It is additionally assumed to be circularly symmetric. $n_j(k)$ will be referred to as a zero mean complex circularly symmetric (ZMCCS) Gaussian random variable denoted as $n_j(k) \sim \mathcal{CN}(0, \sigma_n^2)$. Moreover, the noise at each receive antenna has the same variance σ_n^2. The noise process is i.i.d. and is white temporally (independence across time) and spatially (independence across antennas): $\mathbf{n}(k) \sim \mathcal{CN}(0, \sigma_n^2 \mathbf{I})$. The noise process $\{\mathbf{n}(k)\}$ is independent from the symbol process $\{\mathbf{x}(k)\}$, meaning that, for all values (k, k'), $\mathbf{x}(k)$ and $\mathbf{n}(k')$ are independent.

The MIMO channel at time k is defined as a matrix $\boldsymbol{H}(k)$ of dimension $M_R \times M_T$:

$$
\boldsymbol{H}(k) = \begin{bmatrix} h_{11}(k) & \cdots & h_{1M_T}(k) \\ \vdots & \ddots & \vdots \\ h_{M_R 1}(k) & \cdots & h_{M_R M_T}(k) \end{bmatrix}
\tag{2.5}
$$

Finally, the discrete-time MIMO input-output relationship is:

$$
\mathbf{y}(k) = \boldsymbol{H}(k)\mathbf{x}(k) + \mathbf{n}(k)
\tag{2.6}
$$

The main assumptions applicable to all channels in the chapter

---

▶ The noise $n_j(k)$ at antenna j is a ZMCCS Gaussian random variable with variance σ_n^2.
▶ The noise process $\{\mathbf{n}(k)\}$ is i.i.d., temporally and spatially white: $\mathbf{n}(k) \sim \mathcal{CN}(0, \sigma_n^2 \mathbf{I})$.
▶ The processes $\{\mathbf{x}(k)\}$ and $\{\mathbf{n}(k)\}$ are assumed to be independent from each other.
▶ CSIR is known at the receiver.

2.5 Capacity of the Time-Invariant Channel

In this section, the channel is assumed *time invariant (TI) and known at the transmitter and the receiver*. First, we examine the fundamental case of a time-invariant SISO channel, also called Additive White Gaussian Noise (AWGN), followed by the case of multiple antennas at the receiver only (SIMO) or multiple antennas at the transmitter only (MISO). Then, we treat the case of MIMO where multiple antennas are present both at the transmitter and the receiver. For each case, we give the capacity as well as the optimal transceiver.

The main assumptions valid throughout this section are as follows.

Main assumptions for the time-invariant MIMO channel

▶ A codeword spans over an asymptotic long data block, thus averaging out the noise.
▶ The channel is time-invariant.
▶ CSIT and CSIR are known.

The average total transmit power is constrained to be smaller than a set value $\bar{P}$. As the channel remains constant and the noise is stationary, there is no reason to have a variable power within a codeword in the optimal transmission strategy. The transmit power from antenna i is $E|x_i|^2$. The sum of transmit powers $\sum_{i=1}^{M_T} E|x_i|^2$ is constrained to be smaller than $\bar{P}$. As $\sum_{i=1}^{M_T} E|x_i|^2 = \text{tr}(R_{xx})$, the transmit power constraint can be written as:

$$\text{tr}(R_{xx}) \leq \bar{P} \qquad (2.7)$$

To maximise the transmission rate, the transmit power is set constant to its maximal value $\bar{P}$.

2.5.1 Capacity of the Time-Invariant SISO Channel

The discrete-time time-invariant SISO channel (see Figure 2.5) is defined by the input to output relationship:

$$y(k) = h\, x(k) + n(k) \qquad (2.8)$$

h is the value of the channel, independent of time. The signal power, which is the power in $h\, x(k)$, is equal to $\bar{P}|h|^2$. The power in the noise $n(k)$ is σ_n^2. The signal to noise ratio (SNR) is defined as the ratio between the signal power and the noise power:

$$\text{SNR}_{\text{AWGN}} = \frac{\bar{P}|h|^2}{\sigma_n^2} \qquad (2.9)$$

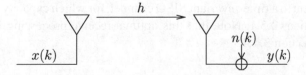

Figure 2.5   Time-invariant SISO channel.

For an AWGN channel, the SNR is essential as it determines the performance of the system. For MIMO systems, the correlation between SISO links plays also an important role.

The capacity for an AWGN channel is equal to:

$$C_{\mathrm{AWGN}}(\mathrm{SNR}_{\mathrm{AWGN}}) = \log_2(1 + \mathrm{SNR}_{\mathrm{AWGN}}) \quad \text{(bits per transmission)} \qquad (2.10)$$

Capacity is achieved when the channel input process $\{x(k)\}$ is a i.i.d. ZMCCS Gaussian process: $x(k) \sim \mathcal{CN}(0, \bar{P})$.

Here, we state some properties of capacity. Note that they are actually not specific to the AWGN channel.

▶ The unit of capacity is *bits per transmission*: it is the number of bits that are transmitted per symbol transmission time T_S.
▶ The capacity can also be defined as $C_{\mathrm{AWGN}}(\mathrm{SNR}_{\mathrm{AWGN}})/T_S$ with unit bits per second.
▶ *Spectral efficiency:* an alternative performance measure to capacity is the spectral efficiency expressed in bit per second per Hz (bits/s/Hz). It is equal to the capacity per second normalised by the system bandwidth W:

$$S_{\mathrm{AWGN}}(\mathrm{SNR}_{\mathrm{AWGN}}) = \frac{C_{\mathrm{AWGN}}(\mathrm{SNR}_{\mathrm{AWGN}})}{T_S W} \quad \text{(bits/s/Hz)} \qquad (2.11)$$

If $T_S W = 1$, $C_{\mathrm{AWGN}}(\mathrm{SNR}_{\mathrm{AWGN}})$ is also equal to the spectral efficiency.
▶ We recall that the quantities involved are all complex: the capacity by real dimension is $\frac{1}{2}\log_2(1 + \mathrm{SNR}_{\mathrm{AWGN}})$.

2.5.2 Time-Invariant SIMO Channel

In a SIMO system, one input symbol $x(k)$ is sent at time k from a single antenna and received at M_R antennas: see Figure 2.6. To better understand the capacity results and how they relate to the SISO channel, we describe the structure of the optimal receiver achieving capacity. This receiver involves a coherent combining of the signals from the multiple antennas based on the CSIR. When including this post-processing, the system becomes equivalent to a time-invariant SISO channel, for which capacity results have been presented in Section 2.5.1. Note that this optimal receive processing is described with more details in Section 3.1.4.

System Model: The SIMO channel is assumed to be time-invariant. The channel coefficients are denoted as h_j, $j, \ldots, M_R$ and are known at the transmitter. At antenna j, the

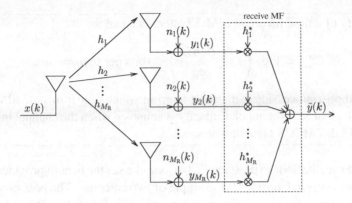

Figure 2.6   SIMO system model with receive beamforming.

received signal $y_j(k)$ is:

$$y_j(k) = h_j x(k) + n_j(k) \tag{2.12}$$

Equivalent SISO Channel: Multiple (distorted) copies of the transmitted symbol are available at the receiver. Each copy carries some information about the transmitted symbol. In order to extract a maximal amount of information from the multiple received signals, a proper processing is necessary. The optimal processing is simple: it is a linear processing called receive spatial matched filtering (MF) or receive maximum ratio combining (MRC), as shown in Figure 2.6. This processing is information lossless. In receive MF, the received signals are aligned in phase so that they can add constructively and a larger weight (scaling coefficient) is assigned to stronger channels. As detailed in Section 3.1.4, MF is the linear processing which maximises the post-processing SNR. The received signal at antenna j is first multiplied by a scalar coefficient h_j^* matched to the channel: the signal is then equal to $|h_j|^2 x(k) + h_j^* n_j(k)$. The processed signals from all antennas are added up, resulting in the processed signal:

$$\tilde{y}(k) = \|\mathbf{h}\|^2 x(k) + \tilde{n}(k) \tag{2.13}$$

where $\mathbf{h} = \begin{bmatrix} h_1 \cdots h_{M_R} \end{bmatrix}^{\mathrm{T}}$ and $\tilde{n}(k) = \sum_{i=1}^{M_R} h_j^* n_i(k)$. The SIMO system becomes equivalent to a scalar AWGN channel for which the results of the previous section can be applied.

Capacity: The system with input $x(k)$ and output $\tilde{y}(k)$ is a scalar AWGN channel with SNR equal to

$$\mathrm{SNR}_{\mathrm{SIMO}} = \frac{\bar{P}\|\mathbf{h}\|^2}{\sigma_n^2} \tag{2.14}$$

> The capacity of a time-invariant SIMO system is equal to:
>
> $$C_{\text{SIMO}}^{\text{TI}} = \log_2 \left(1 + \frac{\bar{P} \|\mathbf{h}\|^2}{\sigma_n^2} \right) \quad \text{(bits per transmission)} \quad (2.15)$$
>
> The optimal post-processing is matched filtering which transforms the SIMO channel into an equivalent SISO channel. Capacity is achieved when the channel input process $\{x(k)\}$ is a i.i.d. ZMCCS Gaussian process.

Array Gain: From the SNR expression (2.14), one can see the benefit provided by multiple antennas at the receiver. Consider the example of two antennas. The post-processing SNR when only the first antenna is active at the receiver is: $\bar{P}|h_1|^2/\sigma_n^2$. When both antennas are active, the SNR is $\bar{P}|h_1|^2/\sigma_n^2 + \bar{P}|h_2|^2/\sigma_n^2$. The postprocessing SNR increases by $\bar{P}|h_2|^2/\sigma_n^2$, representing an *array gain*.

2.5.3   Time-Invariant MISO Channel

A MISO system consists of several antennas at the transmitter and a single antenna at the receiver. When the channel is known at the transmitter, an optimal pre-processing performed on the transmitted signals transforms the MISO system into an equivalent SISO channel. This pre-processing is information lossless and achieves capacity. In this scheme, a single symbol is sent at each transmission time. As there are several antennas at the transmitter available to send this single symbol, the question is how to optimally use this multiplicity of antennas to get optimal performance. From each antenna, a signal is sent that is a simple scaled version of the transmitted symbol. At the receiver, all the signals are added up. The pre-processing accounts for the value of the MISO channel to insure that the signals are added up constructively at the receiver.

System Model: In the time-invariant MISO system, the signal $x_i(k)$ is sent from antenna i and received through the time-invariant channel h_i (see Figure 2.7). The channel coefficients are denoted as h_i, $i = 1, \ldots, M_\text{T}$ and are known at the transmitter. The received signal $y(k)$ is:

$$y(k) = \sum_{i=1}^{M_\text{T}} h_i x_i(k) + n(k) = \mathbf{h}^\text{T} \mathbf{x}(k) + n(k) \quad (2.16)$$

where $\mathbf{h} = \begin{bmatrix} h_1 \cdots h_{M_\text{T}} \end{bmatrix}^\text{T}$ and $\mathbf{x}(k) = \begin{bmatrix} x_1(k) \cdots x_{M_\text{T}}(k) \end{bmatrix}^\text{T}$.

Equivalent SISO Channel: The signals $x_i(k)$ achieving capacity have a special structure resulting from a preprocessing called transmit spatial matched filtering or transmit MRC, which is further detailed in Section 3.2.1. In transmit MF, only one symbol $\tilde{x}(k)$ is actually sent at time k. It goes through a linear pre-processing which depends on the channel: see

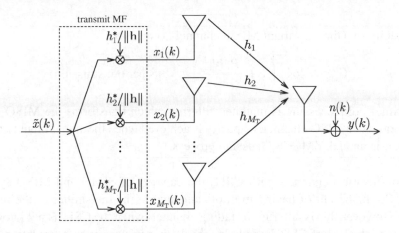

Figure 2.7   MISO system model with transmit beamforming.

Figure 2.7. The transmitted symbol $x_i(k)$ is scaled by the coefficient h_i^* matched to the channel h_i, before being sent from antenna i. We observe that, due to the perfect knowledge of the channel on the transmit side, the weights applied in transmit MF are exactly those that would have been applied in receive MF if the roles of transmitter and receiver had been reversed.

The output of the pre-processing is normalised to comply with the transmit power constraint: $\sum_{i=1}^{M_T} E|x_i(k)|^2 \leq \bar{P}$. Assuming that the transmit power for the signal $\tilde{x}(k)$ is $\bar{P}$, the signal sent from antenna i complying with the power constraint is:

$$x_i(k) = \frac{h_i^*}{\|\mathbf{h}\|}\tilde{x}(k) \tag{2.17}$$

At the receive antenna, all the signals are phase aligned and are added constructively, as for receive MF, and more power is allocated to the stronger channels. Transmit MF maximises the post-processing. The system with input $\tilde{x}(k)$ and output $y(k)$ is a scalar AWGN channel:

$$y(k) = \|\mathbf{h}\|\tilde{x}(k) + n(k) \tag{2.18}$$

Capacity: The SNR at the receive antenna is equal to:

$$\mathrm{SNR}_{\mathrm{MISO}}^{\mathrm{TI}} = \frac{\bar{P}\|\mathbf{h}\|^2}{\sigma_n^2} \tag{2.19}$$

> The capacity of a time-invariant MISO channel is equal to:
>
> $$C_{\text{MISO}}^{\text{TI}} = \log_2 \left(1 + \frac{\bar{P}\|\mathbf{h}\|^2}{\sigma_n^2} \right) \quad \text{(bits per transmission)} \qquad (2.20)$$
>
> The optimal pre-processing is matched filtering which transforms the MISO channel into an equivalent SISO channel. Capacity is achieved when the signal input to transmit MF $\{\tilde{x}(k)\}$ is an i.i.d. ZMCCS Gaussian process.

For time-invariant channels with CSIT, the capacity of SIMO and MISO systems is identical. So the benefit of having multiple antennas at the transmitter or the receiver is the same. However, this is not true for fading channels when the CSIT is not known at the transmitter as the lack of CSIT knowledge results in a performance degradation.

2.5.4   Time-Invariant MIMO Channel: A Set of Parallel Independent AWGN Channels

For the time-invariant SIMO and MISO channels, a single symbol is sent per transmission period. The optimal transceiver involves a transmission or reception relying on a weighting at each transmit or receive antenna. This pre- and post-processing is directly matched to the channel.

When multiple antennas are present simultaneously at both the transmitter and receiver, the capacity achieving scheme consists of sending multiple symbols per transmission period. The transmission and reception of each symbol relies on pre- and post-processing that is matched to the underlying structure of the channel based on its singular value decomposition (SVD). This pre- and post-processing allows for the extraction of a *spatial route* for communication (simply a SISO channel) for each transmitted symbol. Multiple pairs of pre- and post-processing create multiple spatial routes. Those multiple spatial routes are *independent* and the MIMO system becomes equivalent to a set of independent SISO channels. Hence the capacity becomes the sum of the capacity of each SISO channel. In Section 3.2.1, we interpret the aforementioned pre- and post-processing as transmit and receive beamforming.

In summary, when both CSIT and CSIR are available, a SIMO and MISO channel can be made equivalent to a SISO channel. When multiple antennas are present at both the transmitter and the receiver, a MIMO channel can be made equivalent to multiple SISO channels.

2.5.4.1   Singular Value Decomposition of H

The SVD decomposition of the channel matrix is fundamental in understanding MIMO systems: (a) it extracts the equivalent independent AWGN channel structure, (b) it gives the maximum number of streams that can be multiplexed simultaneously, (c) it provides

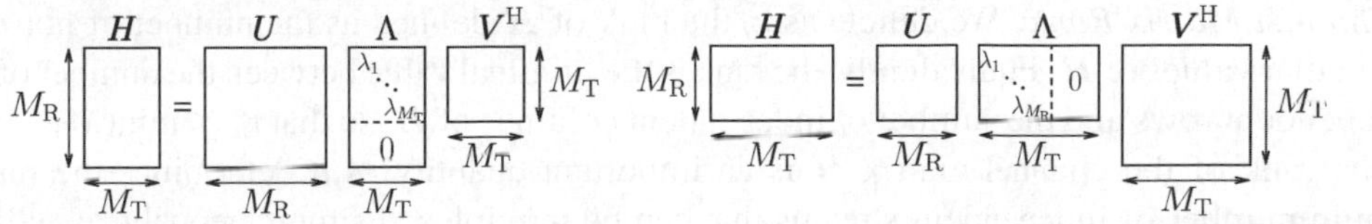

Figure 2.8 Singular value decomposition of the channel matrix H for the case $M_T \leq M_R$ (left) and $M_T \geq M_R$ (right).

a very simple way to compute the capacity which becomes the sum of AWGN channel capacity. The SVD of the channel matrix H is:

$$H = U\Lambda V^H \tag{2.21}$$

The SVD is illustrated in Figure 2.8 for the case $M_T \leq M_R$ and the case $M_T \geq M_R$. Both $M_R \times M_R$ matrix U and $M_T \times M_T$ matrix V are unitary matrices. Λ is a $M_R \times M_T$ diagonal matrix with nonnegative singular values λ_k, $k = 1, \ldots, M_{\min}$, where $M_{\min} = \min(M_T, M_R)$. For convenience, the singular values are ordered decreasingly: $\lambda_1 \geq \lambda_2 \geq \cdots \geq \lambda_{M_{\min}}$. The λ_k's are called the *eigenmodes* of the channel.

A 2×2 system as an example. To illustrate the SVD, we give the example of a 2×2 MIMO system. The channel matrix is written as:

$$H = \begin{bmatrix} h_{11} & h_{12} \\ h_{21} & h_{22} \end{bmatrix} \tag{2.22}$$

Its singular value decomposition is:

$$H = \begin{bmatrix} h_{11} & h_{12} \\ h_{21} & h_{22} \end{bmatrix} = U\Lambda V^H = \begin{bmatrix} \mathbf{u}_1 & \mathbf{u}_2 \end{bmatrix} \begin{bmatrix} \lambda_1 & 0 \\ 0 & \lambda_2 \end{bmatrix} \begin{bmatrix} \mathbf{v}_1^H \\ \mathbf{v}_2^H \end{bmatrix} \tag{2.23}$$

where

$$\mathbf{v}_1 = \begin{bmatrix} v_{11} \\ v_{12} \end{bmatrix}, \ \mathbf{v}_2 = \begin{bmatrix} v_{21} \\ v_{22} \end{bmatrix}, \ \mathbf{u}_1 = \begin{bmatrix} u_{11} \\ u_{12} \end{bmatrix}, \ \mathbf{u}_2 = \begin{bmatrix} v_{21} \\ u_{22} \end{bmatrix}. \tag{2.24}$$

Moreover, the orthogonality constraint for the eigenvectors implies that:

$$U^H U = \begin{bmatrix} 1 & 0 \\ 0 & 1 \end{bmatrix} \ \Leftrightarrow \ \mathbf{u}_1{}^H \mathbf{u}_1 = 1, \ \mathbf{u}_2{}^H \mathbf{u}_2 = 1, \ \mathbf{u}_1{}^H \mathbf{u}_2 = 0 \tag{2.25}$$

$$\Leftrightarrow \ |u_{11}|^2 + |u_{12}|^2 = 1, \ |u_{21}|^2 + |u_{22}|^2 = 1, \tag{2.26}$$

$$u_{11}^* u_{21} + u_{21}^* u_{22} = 0. \tag{2.27}$$

A similar constraint applies for the matrix V.

Channel Matrix Rank: We denote as r_H the rank of H defined as the number of nonzero singular values of H. Equivalently, the rank is the minimal value between the number of independent rows and the number of independent columns of H, so that $r_H \leq \min(M_T, M_R)$. The rank of the channel matrix H is an important quantity as it determines the maximum number of independent streams that can be multiplexed simultaneously, as will be explained later.

Singular Values and Channel Energy: The following relationship between channel energy and singular values will be useful. Indeed, the energy in the channel $\sum_{i=1}^{M_T} \sum_{j=1}^{M_R} |h_{ji}|^2$ can be rewritten as $\mathrm{tr}(HH^H)$. Using the relationships $HH^H = U\Lambda^2 U^H$ and $\mathrm{tr}(HH^H) = \mathrm{tr}(H^H H)$, the following result is obtained: $\mathrm{tr}(HH^H) = \sum_{k=1}^{r_H} \lambda_k^2$. To summarise:

$$\mathrm{tr}(HH^H) = \sum_{i=1}^{M_T} \sum_{j=1}^{M_R} |h_{ji}|^2 = \sum_{k=1}^{r_H} \lambda_i^2 \qquad (2.28)$$

2.5.4.2 Equivalent MIMO System

We proceed to the following change of variable:

$$\begin{cases} \tilde{x} = V^H x \\ \tilde{y} = U^H y \end{cases} \quad \Longleftrightarrow \quad \begin{cases} x = V\tilde{x} \\ y = U\tilde{y} \end{cases} \qquad (2.29)$$

Note that, for clarity reasons, we give up the time indices in all quantities. As U and V are invertible, x and y can be recovered uniquely from $\tilde{x}$ and $\tilde{y}$. Hence, the new system with input $\tilde{x}$ and outputs $\tilde{y}$ is an equivalent representation of the original system. This equivalent system is the result of a linear pre-processing at the transmitter and a linear post-processing at the receiver: see Figure 2.9. Let us now examine the details of this processing.

2.5.4.3 Multi Pre and Post Processing

Going back to the 2×2 MIMO example, the 2×1 transmitted vector x can be rewritten as:

$$x = v_1 \tilde{x}_1 + v_2 \tilde{x}_2 \qquad (2.30)$$

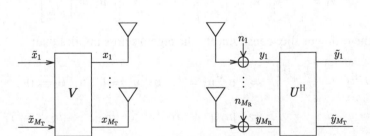

Figure 2.9   Equivalent MIMO model with pre-processing V and post-processing U^H.

The vectors $\mathbf{v}_i$ are transmit pre-processing vectors and are matched to the conjugate of the right singular vectors of H. After being transferred through the channel, the signal is transformed as $H\mathbf{x} = U \mathbf{\Lambda} V^H V \tilde{\mathbf{x}} = U \mathbf{\Lambda} \tilde{\mathbf{x}}$. The received signal is:

$$\mathbf{y} = \lambda_1 \mathbf{u}_1 \tilde{x}_1 + \lambda_2 \mathbf{u}_2 \tilde{x}_2 + \mathbf{n}. \qquad (2.31)$$

Each element $\tilde{y}_k$ of $\tilde{\mathbf{y}}$ is obtained by post-processing the received signal $\mathbf{y}$ by $\mathbf{u}_k^H$. Using the orthogonality property of the singular vectors:

$$\tilde{y}_k = \lambda_k \tilde{x}_k + \mathbf{u}_k^H \mathbf{n} \qquad (2.32)$$

Finally, the vectorial input–output relationship of the equivalent system is:

$$\tilde{\mathbf{y}} = \mathbf{\Lambda} \tilde{\mathbf{x}} + \tilde{\mathbf{n}}, \qquad (2.33)$$

where $\tilde{\mathbf{n}} = U^H \mathbf{n}$. Because matrices U and V are unitary, $E(\tilde{\mathbf{x}}\tilde{\mathbf{x}}^H) = E(\mathbf{x}\mathbf{x}^H)$ and $E(\tilde{\mathbf{n}}\tilde{\mathbf{n}}^H) = E(\mathbf{n}\mathbf{n}^H)$. So, the original power constraint on $\mathbf{x}$ still applies to $\tilde{\mathbf{x}}$: $E\|\tilde{\mathbf{x}}\|^2 \leq \bar{P}$. Furthermore, the noise $\tilde{\mathbf{n}} = U^H \mathbf{n}$, as a linear combination of Gaussian variables, keeps the same statistics as $\mathbf{n}$: $\tilde{\mathbf{n}} \sim \mathcal{CN}(0, \sigma_n^2 I)$, namely independent Gaussian random variables of the same variance.

2.5.4.4 Independent AWGN Channels

Each output k, $k = 1, \ldots, M_R$ of the equivalent system (2.33) can be written as:

$$\begin{aligned} \tilde{y}_k &= \lambda_k \tilde{x}_k + \tilde{n}_k, &\quad \text{for} \quad k &= 1, \ldots, r_H \\ \tilde{y}_k &= \tilde{n}_k, &\quad \text{for} \quad k &= r_H + 1, \ldots, M_R, \text{ when } r_H < M_R. \end{aligned} \qquad (2.34)$$

Each independent channel is also called an *eigenchannel* as the associated channel coefficient is an eigenvalue of the channel matrix. Alternatively it is sometimes referred to as a *subchannel*. The entire MIMO channel is equivalent to the set of all the eigenchannels, each of which has a different SNR.

When the rank of the matrix is strictly smaller than M_T, data $\tilde{x}_k, k > r_H$ gets transmitted through an equivalent channel $\lambda_k = 0$: no information about $\tilde{x}_k$ is contained at the receiver. So the maximum number of codewords that can be successfully decoded at the receiver is equal to the rank of the channel matrix. Furthermore, only the first r_H signals $\tilde{y}_k$ should be considered for decoding.

Looking at Equation (2.34), each input–output relationship $\tilde{y}_k = \lambda_k \tilde{x}_k + \tilde{n}_k$ describes an AWGN channel as described in Section 2.5.1: see Figure 2.10. Furthermore, as the additive noises $\tilde{n}_k$ are all independent from each other, those AWGN subchannels are all independent from each other, forming a set of parallel AWGN channels. This means that an optimal coding can be done independently for each AWGN subchannel. Thus, the capacity of the MIMO system is the sum of the individual capacities.

The change of variable in Equation (2.29) brings out a structure with parallel eigenchannels. This structure allows for a simple derivation of the MIMO capacity as the sum

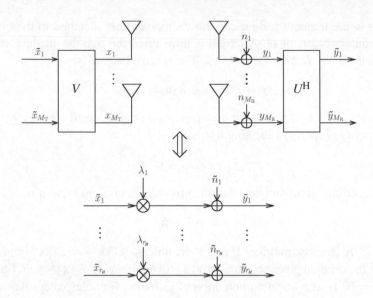

Figure 2.10 Equivalent MIMO model consisting of r_H AWGN independent channels.

of the capacity of each eigenchannel. One element is still missing to completely determine the capacity, namely the transmit power allocated to each eigenchannel. This is what is determined next.

2.5.4.5 Waterfilling and Capacity

Let us introduce the following quantities defined for each eigenchannel as:

$$\gamma_k = \frac{\lambda_k^2}{\sigma_n^2}, \quad k = 1, \ldots, r_H \tag{2.35}$$

Let P_k be the transmit power of eigenchannel k (the power in $\tilde{x}_k$). $P_k\gamma_k$ can be seen as the SNR of the kth eigenchannel. The capacity of each eigenchannel with transmit power P_k is the capacity of an AWGN channel with SNR $= P_k\gamma_k$: it is equal to $\log_2(1 + P_k\gamma_k)$. P_k is adjusted to maximise the capacity of the MIMO system while complying with the overall transmit power constraint: $\sum_{i=1}^{r_H} P_k \leq \bar{P}$. In general, P_k depends on all nonzero singular values, through the power constraint.

The capacity of the MIMO system is the sum of the individual capacities with optimised transmit power per eigenchannel:

$$C_{\text{MIMO}}^{\text{TI}} = \max_{P_k \text{ s.t. } \sum_{k=1}^{r_H} P_k \leq \bar{P}} \sum_{i=1}^{r_H} \log_2\left(1 + P_k\gamma_k\right). \tag{2.36}$$

Note that MIMO capacity only depends on the γ_k's and hence the singular values of the channel matrix.

The optimisation problem in (2.36) can be solved using the method of Lagrangian multipliers. The results on capacity can be summarised as follows.

The capacity of the time-invariant MIMO channel is:

$$C_{\text{MIMO}}^{\text{TI}} = \sum_{k=1}^{r_H} \log_2 \left(1 + P_k^o \gamma_k\right) \tag{2.37}$$

where the transmit power is allocated across antennas as:

$$P_k^o = \left(\frac{1}{\gamma_0} - \frac{1}{\gamma_k}\right)^+ \tag{2.38}$$

γ_0 is the cut-off value and is determined using the power constraint :

$$\sum_{k=1}^{r_H} P_k^o = \sum_{k=1}^{r_H} \left(\frac{1}{\gamma_0} - \frac{1}{\gamma_k}\right)^+ = \bar{P} \tag{2.39}$$

An optimal transceiver, called *eigenmode transmission*, is depicted in Figure 2.11 where:

- r_H independently coded AWGN codewords are assigned power P_k^o;
- they are subject to a linear processing by matrix V and sent through the M_T transmit antennas;
- M_R received signals are subject to a linear processing by matrix U^H; the r_H resulting signals are decoded using an AWGN decoder;
- the optimal input vector $\tilde{\mathbf{x}}$ is a i.i.d. ZMCCS Gaussian random variable.

This transceiver structure is not unique and there exist other structures achieving capacity.

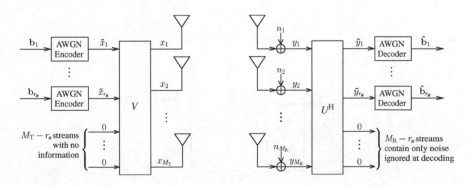

Figure 2.11   General structure of eigenmode transmission, achieving capacity.

2.5.4.6 Waterfilling: a Numerical Example

Consider a 4×4 MIMO system with singular values: $\lambda_1^2 = 3.2$, $\lambda_2^2 = 1.2$, $\lambda_3^2 = 0.44$, $\lambda_4^2 = 0.24$. The SNR ρ defined as $\rho = \bar{P}/\sigma_n^2$ is fixed at 5 dB. $\bar{P}$ is fixed to 1 so that $\sigma_n^2 \approx 0.3162$. In the table below, the different steps leading to the optimal waterfilling solution are described.

Eigenchannel number	1	2	3	4
Singular values λ_k^2	3.2	1.2	0.44	0.24
$1/\gamma_k = \sigma_n^2/\lambda_k^2$	0.0988	0.2635	0.7187	1.3176
$1/\gamma_0 = (\bar{P}+\sum_{k=1}^{4} 1/\gamma_k)/4 = 0.8497$				
Test if $1/\gamma_0 - 1/\gamma_k > 0$	Yes	Yes	Yes	No $\Rightarrow P_4^o = 0$
$1/\gamma_0 = \bar{P} + \sum_{k=1}^{3} 1/\gamma_i = 0.6937$				
Test if $1/\gamma_0 - 1/\gamma_i > 0$	Yes	Yes	No	$\times$ $\Rightarrow P_3^o = 0$
$1/\gamma_0 = \bar{P} + \sum_{k=1}^{2} 1/\gamma_k = 0.6812$				
Test if $1/\gamma_0 - 1/\gamma_k > 0$	Yes	Yes	$\times$	$\times$
Final power allocation P_k^o	0.5824	0.4176	0	0

Figure 2.12 shows the values $1/\gamma_k$ of each eigenchannel (light grey zone) as well as the cut-off value $1/\gamma_o$. No power is assigned in eigenchannel k if $1/\gamma_k > 1/\gamma_o$. The darker grey zone shows the power level. This power allocation is called waterfilling: considering the light grey zones as a container, power allocation is similar to pouring water into this container until the total power is assigned. Power is first allocated to the first eigenchannel, until the water reaches the level $1/\gamma_2$ at which point power starts to be allocated to eigenchannel 2 as well. In our example, the maximal power is allocated ($P_1^o + P_2^o = \bar{P}$) before reaching the level $1/\gamma_3$, so eigenchannel 3 and 4 are not assigned any power.

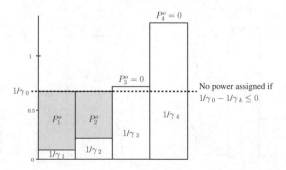

Figure 2.12 Waterfilling: numerical example. Water is poured until the available power is fully assigned.

2.5.4.7 Performance of Eigenmode Transmission

Capacity is used to characterise the performance of a time-invariant MIMO system. Performance is presented according to the following SNR, defined as the ratio between the total transmit power and the noise variance at each receive antenna:

$$\rho = \frac{\bar{P}}{\sigma_n^2} \qquad (2.40)$$

We describe the capacity behaviour at high and low SNR, ρ where performance is linked to the behaviour of the waterfilling power allocation.

Waterfilling at low and high SNR: Figure 2.13 presents an example of waterfilling in the high and low SNR regime. The channel has four nonzero eigenvalues (from the numerical example in Section 2.5.4.6). We recall the optimal power allocation:

$$P_k^o = \left(\frac{1}{\gamma_0} - \frac{1}{\gamma_k} \right)^+ \qquad (2.41)$$

- At high SNR, the values $1/\gamma_k$ are small compared to $1/\gamma_0$. So the power allocated to each eigenchannel is approximately constant and equal to $\bar{P}/r_H$.
- At low SNR, the values $1/\gamma_k$ are large compared to $1/\gamma_0$. Only the eigenchannel with highest eigenvalue is allocated with the whole available power.

Performance at high SNR: At high SNR, the optimal waterfilling policy allocates the same power $\bar{P}/r_H$ to the nonzero eigenmodes. Hence, capacity becomes:

$$C_{\text{MIMO}}^{\text{TI}} \approx \sum_{k=1}^{r_H} \log_2 \left(1 + \frac{\bar{P}}{\sigma_n^2} \frac{\lambda_k^2}{r_H} \right) \qquad (2.42)$$

At high SNR, performance is mostly determined by the rank of the channel matrix and the MIMO channel conditioning.

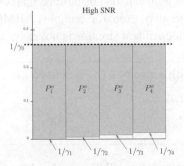
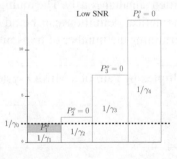

Figure 2.13 Waterfilling example at high and low SNR. At high SNR (left), all the eigenchannels are assigned approximately the same power. At low SNR (right), only the eigenchannel with highest SNR is assigned power.

- *Channel matrix rank*: From the high SNR approximation of capacity (2.42), we obtain:

$$C^{TI}_{MIMO} \approx r_H \log_2(\rho) \tag{2.43}$$

At high SNR, capacity scales linearly with r_H and is about r_H times larger than the capacity of an AWGN channel.

- *MIMO channel conditioning*: Consider all the MIMO channels with same energy, that is, same value of $\sum_{i=1}^{r_H} \lambda_i^2$. For such channels and at high SNR, one can show that capacity (2.42) is maximised when all eigenvalues are equal. This result is important as it reveals a property of MIMO channel crucial to support multiplexing and reach high data rates at high SNR. A useful measure to quantify the performance of a MIMO channel is the condition number defined as: $\max_k(|\lambda_k|)/\min_k(|\lambda_k|)$. When the condition number is close to 1, the MIMO channel is said to be well-conditioned and supports a high multiplexed data rate.

Performance at low SNR: At low SNR, the optimal waterfilling policy allocates all the power to the strongest eigenmode. Hence, capacity becomes:

$$C^{TI}_{MIMO} \approx \log_2\left(1 + \rho\lambda^2_{max}\right) \tag{2.44}$$

At low SNR, the optimal strategy does not rely on multiplexing. It consists in sending a single stream through a channel with highest possible energy, which is guaranteed by proper preprocessing and postprocessing. An appropriate combination of transmit signals and receive signals based on CSI is necessary, resulting in a maximal array gain both at the transmitter and the receiver.

2.5.4.8   Definition of Multiplexing Gain

The multiplexing gain has been defined as the number of independent streams that can be transmitted simultaneously. The multiplexing gain can also be defined mathematically. This mathematical definition can be advantageously used for complex MIMO system when determining the number of transmitted independent streams is not easy.

The multiplexing gain of a MIMO system with capacity $C(\rho)$ is defined as:

$$\lim_{\rho\to\infty} \frac{C(\rho)}{\log_2 \rho} \tag{2.45}$$

This definition makes sense as multiplexing occurs at high SNR only. Looking at the high SNR capacity approximation in (2.43) and applying this definition, we observe that the multiplexing gain is indeed equal to the number of independent multiplexed streams r_H.

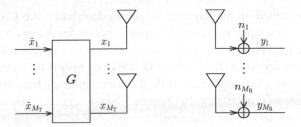

Figure 2.14   MIMO System with fixed input covariance matrix $R_{xx} = GG^H$.

2.5.5   Maximal Achievable Rate for Fixed Input Covariance Matrix

Let us examine the MIMO architecture in Figure 2.14, with assumptions:

- The inputs $\tilde{x}_i$ are independent across antennas.
- The vector input $\tilde{x}$ is pre-processed by a matrix G as $x = G\tilde{x}$. The covariance matrix of the input vector is $Exx^H = R_{xx} = GG^H$.

This architecture corresponds to eigenmode transmission when the pre-processing matrix G is equal to $V\text{diag}(P_1^o, \ldots, P_{M_T}^o)^{1/2}$, where V comes from the SVD of the channel matrix (2.21) and P_i^o comes from the waterfilling power allocation (2.41).

There are situations where such a pre-processing is not possible. When the computational complexity at the transmitter is limited, it is simpler to send independent codewords directly from each antenna, in which case $G = I$. It is also simpler to pre-process the data by a matrix that does not depend on the channel, so that the pre-processing does not change each time the channel changes.

Such an architecture is actually common and the question is how to characterise its performance. This section treats the case where *the channel coefficients are known at the transmitter* and yet the optimal capacity achieving pre-processing is not performed.

2.5.5.1   System Performance

The performance of the system is based on the following result.

Consider a MIMO system $y = Hx + n$ under the following assumptions: (1) the channel is time invariant and known at the transmitter, (2) the input x has a fixed covariance matrix R_{xx}. For such a system with a fixed input covariance matrix, the maximal rate for reliable communication is:

$$\mathcal{R}_{max} = \log_2 \det \left[I + \frac{1}{\sigma_n^2} H R_{xx} H^H \right] \tag{2.46}$$

This maximal rate is achieved when the input vector has a ZMCCS Gaussian distribution.

(2.46) can be interpreted as the capacity of the system when the CSI is known at the transmitter and the covariance matrix of the input signals is fixed. However, it is not labelled as a capacity in general but rather a maximal achievable rate for reliable communication.

In general, as the preprocessing matrix G does not match the channel realisation, a simple linear processing is not optimal while it is for eigenmode transmission. A ML joint decoding of the streams is necessary. In Chapter 3, a simpler receiver structure is presented based on successive interference cancellation.

2.5.5.2   Reformulating the Capacity for the Time-Invariant MIMO Channel

Let us now revisit the time-invariant MIMO channel of Section 2.5.4. For a fixed covariance matrix, the maximal achievable rate for reliable communication is (2.46). To find the capacity, one needs to find the optimal covariance matrix that maximises (2.46). The capacity is then also defined as:

$$C_{\text{MIMO}}^{\text{TI}} = \max_{R_{xx}, \text{tr}(R_{xx}) \leq \bar{P}} \log_2 \det \left[I + \frac{1}{\sigma_n^2} H R_{xx} H^{\text{H}} \right] \tag{2.47}$$

The optimal correlation matrix has the expression $R_{xx}^{\text{o}} = V \text{diag}(P_1^{\text{o}}, \ldots, P_{M_\text{T}}^{\text{o}}) V^{\text{H}}$ for which capacity can be written as (2.37).

Capacity for time-invariant channels

For time-invariant channels, the channel coefficients are assumed to be known at both the transmitter and the receiver. The noise variance is also known at the transmitter. The total transmit power is limited to $\bar{P}$.

▶ AWGN channel $y = hx + n$.
The capacity is:

$$C_{\text{AWGN}} = \log_2 \left(1 + \frac{\bar{P}|h|^2}{\sigma_n^2} \right) \tag{2.48}$$

The capacity achieving code has a constant power $\bar{P}$ and constant rate (equal to the capacity). The optimal input signal has a zero mean Gaussian circular symmetric complex distribution.

▶ SIMO channel $\mathbf{y} = \mathbf{h}x + \mathbf{n}$ and MISO channel $y = \mathbf{h}^{\text{T}}\mathbf{x} + n$

Both channels are equivalent to an AWGN channel. The capacity achieving structure involves a linear processing matched to the channel.

– SIMO: the received signals are processed as: $\tilde{y} = \mathbf{h}^{\text{H}}\mathbf{y}$
– MISO: symbol $\tilde{\mathbf{x}}$ is pre-processed as $\mathbf{x} = \mathbf{h}^*/\|\mathbf{h}\| \tilde{x}$ before being transmitted.

> **Capacity for time-invariant channels (Continued)**

For both SIMO and MISO channels, the equivalent SISO channel has an SNR equal to $\bar{P}\|\mathbf{h}\|^2/\sigma_n^2$. The SIMO and MISO channel have the same capacity expression:

$$C_{\text{SIMO/MISO}}^{\text{TI}} = \log_2\left(1 + \frac{\bar{P}\|\mathbf{h}\|^2}{\sigma_n^2}\right) \tag{2.49}$$

If $M_T = M_R$, both SIMO and MISO have the same capacity. Compared to a SISO channel, a SIMO or MISO channel exhibit an array gain (M_R for SIMO and M_T for MISO) due to the presence of multiple antennas at one end of communication link.

▶ MIMO channel $\mathbf{y} = H\mathbf{x} + \mathbf{n}$.

The major advantage of a MIMO system over a SIMO/MISO system is a throughput boost thanks to its ability to create multiple independent spatial AWGN channels. SIMO and MISO systems create a single spatial channel.

Those multiple independent spatial channels are created by a linear processing at the transmitter and receiver, based on the SVD of the channel $H = U\Lambda V^H$.

The transmit and receive processing depend on the singular eigenvectors of the channel matrix. The transmit processing is matched to V^H: the transmitted signal $\mathbf{x}$ is pre-processed as $\mathbf{x} = V\tilde{\mathbf{x}}$. The receive processing is matched to U: the received signal $\mathbf{y}$ is processed as $\tilde{\mathbf{y}} = U^H\mathbf{y}$.

The transmit power (power in $\tilde{\mathbf{x}}$) is distributed optimally among the spatial channels. The power allocation is based on the singular values of H and follows the waterfilling policy.

The capacity is the sum of capacities of the individual independent spatial channels:

$$C_{\text{MIMO}}^{\text{TI}} = \sum_{i=1}^{r_H} \log_2\left(1 + \frac{P_i^o|\lambda_i|^2}{\sigma_n^2}\right) \tag{2.50}$$

The number of independent channels is limited by the rank of the channel matrix H. Depending on the total amount of available power, some eigenchannels might not be allocated any power, resulting in a number of active streams smaller than r_H.

Encoding: the active eigenchannels are encoded independently by an AWGN encoder before linear pre-processing. No data is sent on the remaining eigenmodes.

Decoding: after linear post-processing, an independent decoding of the active eigenchannels is performed.

Performance: the main properties of MIMO systems at low and high SNR regimes are as follows.

– At high SNR: MIMO systems exhibit a spatial multiplexing gain equal to r_H. Capacity scales linearly with r_H and is about r_H times larger than the capacity of

Capacity for time-invariant channels (Continued)

an AWGN channel. Performance is influenced by the energy distribution among the eigenvalues. For a fixed channel energy, channels with equal eigenvalues maximise capacity.

– At low SNR: MIMO systems lose their spatial multiplexing capability and can transmit successfully only a single data stream. MIMO systems benefit from array gain at the receiver and/or at the transmitter due to the presence of multiple antennas.

▶ When the covariance matrix R_{xx} of the input signals is fixed, the maximal achievable rate for a reliable communication is:

$$\mathcal{R}_{\max} = \log_2 \det \left[I + \frac{1}{\sigma_n^2} H R_{xx} H^{\mathrm{H}} \right] \tag{2.51}$$

This rate is achieved when the input vector has a *zero-mean complex circular-symmetric Gaussian* distribution.

2.6 Fast Fading Channels with CSIT Distribution: Ergodic Capacity

We consider a narrowband MIMO channel as described in Section (2.4) with vectorial output at time k:

$$\mathbf{y}(k) = H(k)\mathbf{x}(k) + \mathbf{n}(k). \tag{2.52}$$

The channel is a *random variable* following a certain distribution. $H(k)$ is the channel realisation at time k. $\{H(k), k \in \mathbb{N}\}$ is the fading channel process assumed *stationary* and *ergodic*. The fading and noise processes are assumed to be independent. The CSIT comprises the channel and the noise variance at the receive antennas. The coding delay is assumed to be large compared to the channel coherence time: codewords span an asymptotically large number of channel fades and noise samples. Hence, both impairments are averaged out allowing for a reliable communication (see Section 2.2.2).

To summarise, the main assumptions valid throughout this section are as follows.

Main assumptions for the MIMO fast fading channel with distribution of CSIT.

▶ Coding delay is large compared to the channel coherence time. A codeword spans over many fades.
▶ The channel random process is stationary and ergodic. The fading process and noise process are assumed to be independent.
▶ Distribution of CSIT is known at transmitter. Instantaneous CSIT not known. CSIR is known.

The transmit power is constrained to be smaller, in average, than $\bar{P}$, where the average is taken over time and hence over the channel fades. The transmit power constraint is (see Section 2.5):

$$\sum_{i=1}^{M_T} E\left[|x_i|^2\right] = \text{tr}(\boldsymbol{R}_{xx}) \leq \bar{P} \tag{2.53}$$

2.6.1   Ergodic Capacity: Basic Principles

Even if the instantaneous value of the CSIT is not known, a reliable communication can be achieved by coding over an asymptotically large number of channel coherence periods, thus averaging out the channel fades. Based on the CSIT distribution, a maximal rate for reliable communication can be determined. This rate is *constant* over the channel fades and is the *ergodic capacity*.

Transmission cannot be adapted to the channel variations as the instantaneous CSIT is not known. However, *transmission can be adapted to the distribution of the channel*. As the fading process is stationary, the channel distribution characteristics remain the same in time, so the transmit parameters remain constant in time. Therefore, it is optimal to keep the transmit power constant in time and equal to its maximal value $\bar{P}$. For a transmitter with multiple antennas, to get optimal performance, the transmitter should adapt to the spatial distribution of the channel coefficients across antennas. This is done by optimising the spatial distribution of the transmitted signals to the spatial distribution of the channel. In particular, the covariance matrix of the input signals should fit the channel distribution.

Next, we present the ergodic capacity starting from the SISO fading channel, then the SIMO and MISO channels and at last the MIMO channel.

2.6.2   Fast Fading SISO Channel with CSIT Distribution

The input–output relationship of a narrowband SISO channel at time k is:

$$y(k) = h(k)x(k) + n(k) \tag{2.54}$$

The system is subject to the average transmit power constraint $E|x(k)|^2 \leq \bar{P}$. To obtain optimal performance, the transmit power is set to its maximal value $\bar{P}$. Thus, the instantaneous SNR at time k is equal to:

$$\text{SNR}(k) = \frac{\bar{P}|h(k)|^2}{\sigma_n^2} \tag{2.55}$$

The ergodic capacity for a fast fading SISO channel with CSIT distribution is defined as:

$$C_{SISO}^{FF} = E\left[\log_2\left(1 + \frac{\bar{P}|\mathbf{h}|^2}{\sigma_n^2}\right)\right] \quad \text{(bits per transmission)} \qquad (2.56)$$

The capacity achieving coding scheme has a fixed rate and the codewords spread over an asymptotically large number of coherence periods. Note that $\log_2\left(1 + \bar{P}|h(k)|^2/\sigma_n^2\right)$ is the capacity of the channel at time k when the instantaneous value of the CSIT is available. The ergodic capacity is the average of the capacity with CSIT over the channel fades.

2.6.3   Fast Fading SIMO Channel with CSIT Distribution

The input-output relationship of a flat SIMO channel at time k is:

$$\mathbf{y}(k) = \mathbf{h}(k)x(k) + \mathbf{n}(k) \qquad (2.57)$$

The main idea to determine the capacity of this system is as follows: the optimal receiver transforms the SIMO system into an equivalent SISO system for which the result from previous section can be used.

From Section 2.5.2 where the channel is assumed to be known at the transmitter, we have seen that the optimal receiver is the receive matched filter. This receiver transforms the SIMO channel into an equivalent SISO channel without loss of information. A key observation is that receive matched filtering necessitates the knowledge of the channel at the receiver only, not at the transmitter. Hence, it is also the optimal receiver when the channel is not known at the transmitter.

By applying receive matched filtering, the SIMO channel can be made equivalent to the fast fading SISO channel $\mathbf{h}(k)^H\mathbf{y}(k) = \|\mathbf{h}(k)\|^2 x(k) + \mathbf{h}(k)^H\mathbf{n}(k)$. The capacity results for the SISO fast fading channel can be applied here for an SNR at time k equal to $\text{SNR}(k) = \bar{P}\|\mathbf{h}(k)\|^2/\sigma_n^2$.

The ergodic capacity of a fast fading SIMO channel with CSIT distribution is equal to:

$$C_{SIMO}^{FF} = E\left[\log_2\left(1 + \frac{\bar{P}\|\mathbf{h}\|^2}{\sigma_n^2}\right)\right] \quad \text{(bits per transmission)} \qquad (2.58)$$

As was observed for a time-invariant channel, a fast fading SIMO channel exhibits an array gain thanks to the coherent combining of the M_R received signals.

2.6.4   Fast Fading MISO Channel with CSIT Distribution

In this section, we will just state the capacity expression for an i.i.d. Rayleigh fading channel. The MISO channel can be viewed as a special case of the MIMO channel: more details on MISO channels can be found in Section 2.6.5.

The ergodic capacity of a fast fading MISO i.i.d. Rayleigh channel $y(k) = \mathbf{h}(k)^{\mathrm{T}}\mathbf{x}(k) + n(k)$ with CSIT distribution is equal to:

$$C_{\mathrm{MISO}}^{\mathrm{FF}} = E\left[\log_2\left(1 + \frac{1}{M_{\mathrm{T}}}\frac{\bar{P}\|\mathbf{h}\|^2}{\sigma_n^2}\right)\right] \quad \text{(bits per transmission)} \qquad (2.59)$$

The MISO channel exhibits a performance loss of $1/M_{\mathrm{T}}$ compared to the case where the instantaneous value of the CSIT is known.

2.6.5   Fast Fading MIMO Channel with CSIT Distribution

The input–output relationship for the fast fading narrowband MIMO channel is:

$$\mathbf{y}(k) = \mathbf{H}(k)\mathbf{x}(k) + \mathbf{n}(k) \qquad (2.60)$$

2.6.5.1   Capacity for a General MIMO Channel Distribution

For an optimal transmission, the covariance of the input signals $\mathbf{R}_{\mathrm{xx}}$ should be adapted to the channel distribution. When the covariance matrix of the input signal is fixed, we have seen that the maximal achievable rate over a given channel fade $\mathbf{H}$ is $\log_2 \det[\mathbf{I} + \mathbf{H}\mathbf{R}_{\mathrm{xx}}\mathbf{H}^{\mathrm{H}}/\sigma_n^2]$. It is achieved when the channel is known at the transmitter. The ergodic capacity is the average of the maximal achievable rate over the channel fades.

The capacity of a fast fading MIMO channel with CSIT distribution is:

$$C_{\mathrm{MIMO}}^{\mathrm{FF}} = \max_{\mathbf{R}_{\mathrm{xx}}\ \text{s.t. tr}(\mathbf{R}_{\mathrm{xx}})\leq\bar{P}} E\left[\log_2 \det\left[\mathbf{I} + \frac{1}{\sigma_n^2}\mathbf{H}\mathbf{R}_{\mathrm{xx}}\mathbf{H}^{\mathrm{H}}\right]\right] \quad \text{(bits per transmission)}$$

$$(2.61)$$

A capacity achieving code has a constant rate equal to the capacity and spreads over many channel fades.

The optimal coefficients of the correlation matrix have to be determined to compute the ergodic capacity. A closed-form expression of the correlation matrix is not always possible to find except for some simple cases as the case of an i.i.d. Rayleigh fading channel described in the following paragraph.

2.6.5.2 Capacity for a Rayleigh i.i.d. MIMO Channel

The transmit covariance matrix is easily optimised in the particular case of an i.i.d. Rayleigh MIMO channel. For such a channel, the best strategy is an i.i.d. transmission where the symbols transmitted from different antennas are uncorrelated and are assigned the same transmit power, that is, $R_{xx} = \frac{\bar{P}}{M_T} I$.

For an i.i.d. Rayleigh fast fading MIMO channel, the optimal input correlation matrix is $R_{xx} = \frac{\bar{P}}{M_T} I$ and capacity is:

$$C_{\text{MIMO}}^{\text{FF,Rayleigh}} = E\left[\log_2 \det\left[I + \frac{\bar{P}}{M_T \sigma_n^2} HH^H\right]\right] \qquad (2.62)$$

A Rayleigh channel model assumes a rich scattering environment where the rank of the channel matrix is equal to $M_{\min}$ with probability 1. Therefore, the channel matrix possesses $M_{\min}$ nonzero singular values $\{\lambda_k, k = 1, \ldots, M_{\min}\}$ with probability 1. Based on the SVD of the channel, capacity can be rewritten as:

$$C_{\text{MIMO}}^{\text{FF,Rayleigh}} = \sum_{k=1}^{M_{\min}} E_{\lambda_1,\ldots,\lambda_{M_{\min}}} \log_2\left[1 + \frac{\bar{P}}{M_T} \frac{\lambda_k^2}{\sigma_n^2}\right] \qquad (2.63)$$

The expected value is taken with respect to the joint distribution of the singular values $\{\lambda_k\}$. This alternative capacity expression is useful for analysis purposes as seen in the next section.

2.6.5.3 Performance of Fast Fading i.i.d. Rayleigh MIMO Channels

For fast fading channels, we consider only the case of an i.i.d. Rayleigh channel distribution as the capacity has a closed form expression making the analysis straightforward. For performance analysis, we assume for each SISO link h_{ji} that $E\left[|h_{ji}|^2\right] = 1$. When only transmit antenna i is active, the instantaneous receive SNR at receive antenna j at time k is $\bar{P}|h_{ij}(k)|^2/\sigma_n^2$. Hence, $\rho = \bar{P}/\sigma_n^2$ can be seen as the average SNR per SISO link. Performance is described below mostly as a function of ρ.

Channel Conditioning: The first question is: what are the key properties of Rayleigh fading channels resulting in a high value of the capacity? In general, it is difficult to completely characterise the distribution of a MIMO channel maximising the channel capacity $\sum_{k=1}^{M_{\min}} E\left[\log_2\left[1 + \frac{\bar{P}}{M_T} \frac{\lambda_k^2}{\sigma_n^2}\right]\right]$. However, it is possible to give some ideas about the behaviour of such channels. As we have seen for time invariant MIMO channels, for a fixed channel energy, $\sum_{k=1}^{M_{\min}} \log_2\left[1 + \frac{\bar{P}}{M_T} \frac{\lambda_k^2}{\sigma_n^2}\right]$ is maximised when all the eigenvalues are equal. This gives an indication on how well conditioned fading channels should be: the channel energy should be statistically well balanced among singular values. This type of channel is encountered in a rich scattering environment.

Number of Transmit and Receive Antennas: When $M_T = M_R = M$, the asymptotic behaviour of capacity is as follows:

$$C_{\text{MIMO}}^{\text{FF,Rayleigh}} \sim Mg(\rho) \qquad (2.64)$$

where $g(\rho)$ is a function of ρ that depends on the asymptotic behaviour of HH^H/M. We conclude that capacity grows linearly as a function of the number of antennas. Equation (2.64) has to be compared to the asymptotic value (large M_R) of the capacity of a SIMO channel $C \sim \log_2(M_R\rho)$ which grows linearly with $\log_2 M_R$. This linear growth in M (and not a logarithm growth) has sparkled the huge interest in MIMO systems and their spatial multiplexing ability.

Looking at Figure 2.15, we observe:

- MIMO capacity grows linearly with respect to the number of antennas.
- MIMO capacity is approximately M times larger than SISO capacity.
- For MISO, the performance gain is negligible when the number of transmit antennas increases.

High SNR: Even if the channel realisations are not known at the transmitter, fast fading channels have the same properties at high SNR as time-invariant channels. At high SNR, capacity is approximated as:

$$C_{\text{MIMO}}^{\text{FF,Rayleigh}} \sim M_{\min} \log_2 \rho \qquad (2.65)$$

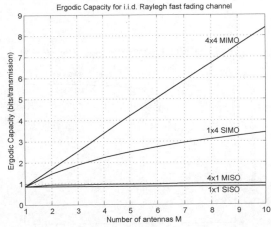

Figure 2.15 Capacity vs number of antennas for a $M \times M$ MIMO system, a $1 \times M$ SIMO system, a $M \times 1$ MISO system and a SISO system.

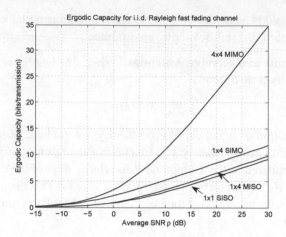

Figure 2.16   Capacity vs SNR ρ.

From the definition in Equation (2.65), we conclude that the multiplexing gain is equal to $M_{\min}$. Capacity grows linearly with $M_{\min}$, logarithmically with the SNR (see Figure 2.16) and is $M_{\min}$ times larger than the AWGN capacity.

Figure 2.17 shows the capacity of a $4 \times M$ and $M \times 4$ MIMO system as the number of antennas M increases at high SNR. As M increases from 1 to 4, the multiplexing gain increases as well from 1 to its maximal value 4. When M becomes larger than 4, the multiplexing gain cannot be increased and performance gains become small. Indeed, at high SNR, performance gains are mostly due to multiplexing gain. Array gain is negligible compared to multiplexing gain. Another illustration of this fact is the MISO case in Figure 2.15, where the capacity increase is negligible when the number of transmit antennas increases.

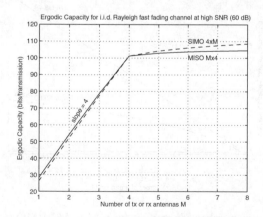

Figure 2.17   Capacity vs number of antenna M for a $4 \times M$ and a $M \times 4$ MIMO system.

Low SNR: Using the relation $\log_2(1 + x) \approx x / \log(2)$ for small x and Equation (2.63), capacity can be approximated as $E\left[\sum_{i=1}^{M_{\min}} \lambda_i^2\right] / \log(2)$. The quantity $E\left[\sum_{i=1}^{M_{\min}} \lambda_i^2\right]$ is the average energy of the channel which is equal to $M_T M_R$ (as $E|h_{ij}|^2 = 1$). Finally, the asymptotic capacity expression is:

$$C_{\text{MIMO}}^{\text{FF,Rayleigh}} \sim M_R \rho / \log 2 \qquad (2.66)$$

(2.66) is also the asymptotic expression for the capacity of a SIMO channel. So, at low SNR, the MIMO system becomes equivalent to a SIMO system (see also Figure 2.16). As for time invariant channels, at high SNR, the best transmission strategy does not rely on stream multiplexing but rather on single stream transmission with array gain at the receiver.

Summary for fast fading channels with CSIR, distribution of CSIT

Even if the CSIT is not available at the transmitter, a reliable communication is possible over fast fading channels with CSIT distribution. Indeed, coding spreads over a very large number of fades, thus averaging out the effects of the channel variations. The performance measure for fast fading channels is the *ergodic capacity*. The capacity achieving codes have a rate that remains constant over the channel fades and equal to the capacity. The transmit power is kept constant over time and equal to its maximum value $\bar{P}$.

▶ SISO/SIMO channel $\mathbf{y}(k) = \mathbf{h}(k)x(k) + \mathbf{n}(k)$.

The optimal receiver is the receive matched filter and capacity is:

$$C_{\text{SIMO}}^{\text{FF}} = E\left[\log_2\left(1 + \frac{\bar{P}\|\mathbf{h}\|^2}{\sigma_n^2}\right)\right] \quad \text{(bits per transmission)} \qquad (2.67)$$

▶ MISO channel $y(k) = \mathbf{h}(k)^{\text{T}}\mathbf{x}(k) + n(k)$.

The MISO fast fading channel is a special case of the MIMO channel. The ergodic capacity is:

$$C_{\text{MISO}}^{\text{FF}} = E\left[\log_2\left(1 + \frac{1}{M_{\text{T}}} \frac{\bar{P}\|\mathbf{h}\|^2}{\sigma_n^2}\right)\right] \quad \text{(bits per transmission)} \qquad (2.68)$$

The MISO channel exhibits a performance loss of $1/M_{\text{T}}$ compared to the case where the CSIT is known.

Summary for fast fading channels with CSIR, distribution of CSIT (Continued)

▶ MIMO channel $\mathbf{y}(k) = \boldsymbol{H}(k)\mathbf{x}(k) + \mathbf{n}(k)$.

The covariance matrix of the input signal $\boldsymbol{R}_{\mathrm{xx}}$ is adapted to the MIMO channel distribution. For a general MIMO channel distribution, capacity is:

$$C_{\mathrm{MIMO}}^{\mathrm{FF}} = \max_{\boldsymbol{R}_{\mathrm{xx}} \text{ s.t. } \mathrm{tr}(\boldsymbol{R}_{\mathrm{xx}}) \leq \bar{P}} E\left[\log_2 \det\left[\boldsymbol{I} + \frac{1}{\sigma_n^2}\boldsymbol{H}\boldsymbol{R}_{\mathrm{xx}}\boldsymbol{H}^{\mathrm{H}}\right]\right] \text{ (bits per transmission)}$$

(2.69)

For an i.i.d. Rayleigh MIMO channel, the optimal covariance matrix is $\boldsymbol{R}_{\mathrm{xx}} = \bar{P}/M_{\mathrm{T}}\boldsymbol{I}$: power is equally distributed among the M_{T} transmit antennas and the signals sent from each antenna are independent from each other. Capacity is:

$$C_{\mathrm{MIMO}}^{\mathrm{FF}} = E\left[\log_2 \det\left[\boldsymbol{I} + \frac{\bar{P}}{M_{\mathrm{T}}\sigma_n^2}\boldsymbol{H}\boldsymbol{H}^{\mathrm{H}}\right]\right] \quad \text{(bits per transmission)} \quad (2.70)$$

The capacity achieving architecture is V-Blast and is described in Chapter 3.

Performance at high and low SNR regime:

- At high SNR: The fast fading MIMO channel exhibits a multiplexing gain of $\min(M_{\mathrm{T}}, M_{\mathrm{R}})$. Capacity scales linearly with $\min(M_{\mathrm{T}}, M_{\mathrm{R}})$ and is about $\min(M_{\mathrm{T}}, M_{\mathrm{R}})$ times larger than the capacity of an AWGN channel.
- At low SNR: MIMO systems lose their spatial multiplexing capability and can transmit successfully only a single data stream. MIMO systems benefit from array gain at the receiver.

2.7 Slow Fading Channel with CSIT Distribution: Outage Probability and Capacity with Outage

In a slow fading channel, coding spreads over a block of data that is much smaller than the coherence time of the channel. Hence, the channel can be approximated as constant over each coding block. The input–output relationship of the slow fading MIMO channel at time k is:

$$\mathbf{y}(k) = \boldsymbol{H}\mathbf{x}(k) + \mathbf{n}(k) \tag{2.71}$$

The channel is a *random variable* denoted as $\mathbf{H}$. The MIMO *channel realisation* during the current block is denoted as $\boldsymbol{H}$. The value of the channel changes from block to block following the fading distribution. The CSIT groups the variance of the noise at the receive antennas and the channel matrix.

Hence, the general assumptions in this section are as follows.

Main assumptions for the MIMO slow fading channel

▶ The coding delay is small compared to the channel coherence time. The channel has a constant value during the duration of a codeword.
▶ The channel random process is stationary and ergodic. The fading process and noise process are assumed to be independent.
▶ Distribution of CSIT is known at transmitter. Instantaneous CSIT is not known. CSIR is known.

As was described for the fast fading channel, the transmit power constraint is:

$$\sum_{i=1}^{M_T} E\left[|x_i|^2\right] = \mathrm{tr}(\boldsymbol{R}_{xx}) \leq \bar{P} \tag{2.72}$$

Likewise, because the fading process is stationary, the transmit power is kept constant in time and equal to its maximal value $\bar{P}$ to maximise performance.

We first outline the principles applied when communicating over slow fading channels. The notion of outage is defined. We also define the notion of antenna diversity which is the main tool to increase performance for slow fading channels with multiple antennas.

2.7.1   Outage: Basic Principles

For a fast fading channel, coding spans over many channel fades, thus averaging out the effect of channel variations. A reliable communication is possible even if the instantaneous value of the channel is not known at the transmitter but only its distribution. For a slow fading channel, the channel remains constant across the codewords. The channel fades cannot be averaged out and it is not possible to guarantee a reliable communication in general.

SISO Example: Let us consider first a SISO channel over a fixed channel fade h. For such a system, the maximal achievable rate for a reliable communication is $\log_2\left(1 + \bar{P}|h|^2/\sigma_n^2\right)$. This maximal rate is the capacity of the system and is achieved when the channel is known at the transmitter. Although we do not make the assumption that the CSIT is known in this section, we use the fact that, if the transmission rate is above capacity, a reliable communication is not possible, whether the CSI is known at the transmitter or not.

When the CSIT is available, the transmitter is able to adjust the transmission rate to be below $\log_2\left(1 + \bar{P}|h|^2/\sigma_n^2\right)$ in order to achieve a reliable communication. In our case, the transmitter cannot rely on the instantaneous value of the CSIT and adjusts the transmission rate based on the CSIT statistics. Assume the transmitter selects a transmission rate R. If R is smaller than $\log_2\left(1 + \bar{P}|h|^2/\sigma_n^2\right)$, then a reliable communication can be achieved. However, in wireless communications, the channel h can take values that are very close to zero (deep fade) and the selected rate R could be larger than the maximal achievable rate $\log_2\left(1 + \bar{P}|h|^2/\sigma_n^2\right)$.

For usual wireless channel distributions, the probability that the selected rate is larger than the capacity $\log_2\left(1 + \bar{P}|h|^2/\sigma_n^2\right)$ is nonzero and could be nonnegligible. Whatever the rate selected by the transmitter, this rate cannot be guaranteed to be smaller than the capacity. Strictly speaking, the capacity for slow fading channels is zero.

MIMO Example: Let us first determine the maximal rate for a reliable communication of a MIMO slow fading channel, over a fixed channel fade $\boldsymbol{H}$. For simplicity reasons, we assume that the transmission is constrained to be i.i.d., meaning that the input covariance matrix is constrained to be equal to $\bar{P}/M_T\boldsymbol{I}$. In fact, the input covariance matrix should be adapted to the channel distribution as explained in Section 2.7.6. When the input covariance matrix is fixed, we know that the maximal transmission rate for reliable communication is equal to:

$$\mathcal{R}_{max} = \log_2 \det\left[\boldsymbol{I} + \frac{\bar{P}}{M_T\sigma_n^2}\boldsymbol{H}\boldsymbol{H}^H\right] \quad \text{(bits per transmission)} \qquad (2.73)$$

This rate is achieved when the instantaneous value of the CSI is known at the receiver. Again, although we do not make this assumption, we use the fact that, if the transmission rate is above $\mathcal{R}_{max}$, a reliable communication is not possible. A transmission rate is selected, based on the distribution of the CSI. Because the channel could be in a deep fade, the probability that the selected rate is larger than $\mathcal{R}_{max}$ is nonzero.

From those examples, we can conclude that a reliable communication cannot be guaranteed in slow fading conditions without CSIT. Therefore, capacity is not a relevant performance measure, and other measures are adopted as described now.

Outage: When the selected transmission rate R is larger than the maximal rate for reliable communication $\mathcal{R}_{max}$ for the given block, the system is said to be in outage. While the outage probability is defined for a targeted transmission rate, the capacity with outage is defined for a targeted outage probability p_o. It is the maximum transmission rate for which the outage probability is smaller than p_o.

For slow fading channels with CSIT distribution, performance is assessed by the *probability of outage* or alternatively the *capacity with outage*.

- For a selected transmit rate R, the outage probability is the probability that R is larger than the maximal achievable rate for reliable communication.
- For a targeted outage probability p_o, the capacity with outage is the maximal rate for which is the outage probability is smaller than p_o.

In a slow fading channel, a transceiver reaches optimal performance when it minimises the outage probability.

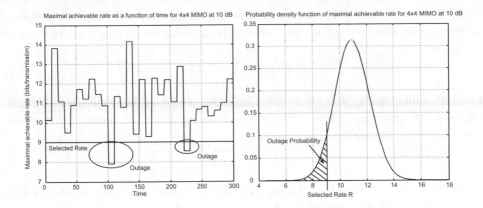

Figure 2.18 Example of a 4×4 MIMO Rayleigh fading channel at $\rho = 10$ dB. Left figure: realisation of the maximal achievable rate $\mathcal{R}_{max}$ for a block fading process. An outage occurs when the selected rate is above the maximal achievable rate. Right figure: probability density function of the maximal achievable rate: outage probability is the probability that the selected rate is above the maximal achievable rate.

This last sentence simply states that if the maximal achievable rate for reliable communication of a given transceiver is smaller than $\mathcal{R}_{max}$, then it will not minimise the outage probability.

Figure 2.18 (left) shows the evolution in time of the maximal achievable rate $\mathcal{R}_{max}$ for a 4×4 MIMO i.i.d. Rayleigh fading channel. According to the channel realisation during the current block, $\mathcal{R}_{max}$ can be large or small. But the transmitter does not know when a favourable or nonfavourable channel realisation happens. For a selected rate of 9 bits per transmission, over the observed period shown in the figure, we observe two outage events. Figure 2.18 (right) shows the probability density function (PDF) of the maximal achievable rate. The outage probability for a transmission rate of 9 bits per transmission is the probability that $\mathcal{R}_{max}$ is smaller than 9: it is the tail of the PDF.

2.7.2 Diversity to Improve Communication Reliability

In multiple antenna communications under slow fading conditions, one of the mechanisms to minimise the outage probability is antenna diversity. In a SISO communication, if the channel is in deep fade, the receiver is not able to decode the signal in general. In a SIMO system, M_R SISO channels are available for communication. The same transmitted signal is sent through the M_R channels. To obtain a gain from diversity, those channels should be independent or have a low correlation factor. Then, when one of the M_R channels is in deep fade, the probability that the other channels are also in a deep fade is low. As the number of independent channels grows, the probability that all the channels are in deep fade decreases. Hence a signal that cannot be decoded in a SISO channel in deep fade is likely to be decoded when it is also sent through other independent channels. A communication with diversity is more reliable. Note that diversity can be also in time and in frequency.

> In a communication relying on diversity, multiple copies of the same signal are sent through multiple independent channels. The diversity gain is the number of independent channels carrying the same signal.

In a MIMO channel, $M_T M_R$ links are available for communication, so the maximal diversity gain is $M_T M_R$. The transmitted signals and the receiver have to be properly designed to extract the full diversity gain. The corresponding transceivers are described in Chapter 3.

Diversity is relied upon specifically in communications over slow fading channels with no CSIT. In time-invariant channels, the transmitter knows the CSIT and adapts to the channel fades by spending less energy in channels in a deep fade. In fast fading channels, codewords span over many channel fades, hence the communication benefits from an asymptotically high (time) diversity gain.

Next, we present the outage results for the SISO/SIMO slow fading channel, then the MISO channel and at last the MIMO channel.

2.7.3   Slow Fading SISO Channels with CSIT Distribution

The input-output relationship of the slow fading SISO channel at time k is:

$$y(k) = h\, x(k) + n(k) \tag{2.74}$$

h is the channel realisation during the current coding block. The SNR is constant over a block and equal to:

$$\text{SNR} = \bar{P}\frac{|h|^2}{\sigma_n^2} \tag{2.75}$$

The maximal rate of this system is the capacity:

$$C_{\text{SISO}}^{\text{TI}} = \log_2\left(1 + \bar{P}\frac{|h|^2}{\sigma_n^2}\right) \tag{2.76}$$

Outage Probability:

For a slow fading SISO channel with CSIT distribution, the outage probability for transmission rate R is defined as the probability that the selected transmission rate R is larger than the capacity:

$$p_{\text{out}}(R) = \Pr\left\{ \log_2\left(1 + \frac{\bar{P}|h|^2}{\sigma_n^2}\right) < R \right\} \tag{2.77}$$

Capacity with Outage: The system decides on a targeted outage probability p_o. Based on the distribution of the CSIT, a transmission rate can be computed such that the associated outage probability is equal to p_o. This transmission rate is the capacity with outage p_o.

For a slow fading SISO channel, the capacity with outage p_o denoted as $C_{\text{out}}(p_o)$ is the transmission rate such that

$$\Pr\left\{ \log_2\left(1 + \frac{\bar{P}|h|^2}{\sigma_n^2}\right) < C_{\text{out}}(p_o) \right\} = p_o \tag{2.78}$$

The codes achieving the capacity with outage p_o are the codes designed for an AWGN channel.

In general, it is difficult to obtain a closed form expression of the capacity with outage. It is possible when a closed form expression for the cumulative distribution function (CDF) of $|h|^2$ exists, or equivalently of $\gamma = |h|^2/\sigma_n^2$. Let us define γ_o such that $\log_2(1 + \bar{P}\gamma_o) = C_{\text{out}}(p_o)$. Then, the outage probability can be written as:

$$\Pr\left\{\log_2(1 + \bar{P}\gamma) < \log_2(1 + \bar{P}\gamma_o)\right\} = p_o \quad \Leftrightarrow \quad \Pr(\gamma < \gamma_o) = p_o \tag{2.79}$$

The right hand side of the equation comes from the fact that $\log_2$ is a strictly monotonic function. $\Pr(\gamma < \gamma_o) = p_o$ can be expressed based on the CDF of γ denoted as $F(\gamma)$:

$$p_o = \Pr(\gamma < \gamma_o) = 1 - F(\gamma_o) \tag{2.80}$$

γ_o can be found by inverting F. Once γ_o is known, $C_{\text{out}}(p_o)$ is determined as $C_{\text{out}}(p_o) = \log_2(1 + \bar{P}\gamma_o)$.

2.7.4   Slow Fading SIMO Channel with CSIT Distribution: Receive Diversity

As the CSIR is assumed to be known, receive MF remains the optimal receiver. It transforms the SIMO channel into an equivalent SISO channel with SNR equal to $\bar{P}\|\mathbf{h}\|^2/\sigma_n^2$. Capacity results for the slow fading SISO channel become applicable.

The outage probability of a slow fading SIMO channel for transmission rate R is defined as:

$$p_{\text{out}}(\text{R}) = \text{Pr}\left\{\log_2\left(1 + \frac{\bar{P}\|\mathbf{h}\|^2}{\sigma_n^2}\right) < \text{R}\right\} \qquad (2.81)$$

A very important difference with the SISO case is that a slow fading SIMO system benefits from *receive diversity* due to the presence of multiple antennas at the receiver. When all channels h_j are independent, the input signal is received through M_R independent paths, thus exhibiting a full diversity gain of M_R. Note that the extraction of full diversity necessitates the optimal processing at the receiver matched to the SIMO channel. When the channels h_j are correlated, the diversity gain decreases.

2.7.5   Slow Fading MISO Channel with CSIT Distribution: Transmit Diversity

To minimise the outage probability, a MISO system relies on *transmit diversity*: multiple copies of the same signal are sent from the multiple antennas. The system should achieve a high level of redundancy at the transmitter while still delivering a high throughput (capacity with outage). The optimal codes for a MISO channel belong to the category of *space–time codes*, codes spanning across time and space. The design of those codes relies solely on the statistics of the CSIT. Further details on space-time codes are given in Chapter 3.

Next, we give the expression of the capacity of a MISO slow fading channel when *transmission is constrained to be i.i.d.* ($\mathbf{R}_{xx} = \bar{P}/M_T\mathbf{I}$). This particular case is important as it provides an upper bound for the performance of space–time block codes (described in Chapter 3). The general case of a transmission that is not constrained to be i.i.d. is a special case of the slow fading MIMO channel treated in Section 2.7.6.

2.7.5.1   Outage Probability for an i.i.d. transmission

For a 2×1 MISO system and when the transmission is constrained to be i.i.d., capacity is achieved using a well-known transmission technique called Alamouti space time block code (STBC). This STBC is described in detail in Chapter 3, Section 3.3. In this section, we state the main results.

By cleverly designing the redundant transmission of two symbols in two transmission periods, the Alamouti STBC extracts the maximal diversity gain while minimising the outage probability. Using the proper post-processing, the MISO system is made equivalent to a SISO channel with SNR:

$$\text{SNR}_{\text{Ala}} = \frac{1}{2} \frac{\bar{P} \|\mathbf{h}\|^2}{\sigma_n^2} \tag{2.82}$$

An important observation comes from the comparison with the MISO channel with CSIT. When the CSIT is available, a transmit matched filter can be applied resulting in the following SNR:

$$\text{SNR}_{\text{TMF}} = \frac{\bar{P} \|\mathbf{h}\|^2}{\sigma_n^2} \tag{2.83}$$

Hence, the penalty for not knowing the instantaneous value of the CSIT is a decrease of the SNR by half.

Based on the SNR (2.82), the maximal achievable rate for reliable communication is $\log_2 \left(1 + \frac{\bar{P}}{2} \frac{\|\mathbf{h}\|^2}{\sigma_n^2} \right)$. Hence, the outage probability of the Alamouti scheme for a transmission rate R is:

$$p_{\text{out}}(R) = \Pr \left\{ \log_2 \left(1 + \frac{\bar{P}}{2} \frac{\|\mathbf{h}\|^2}{\sigma_n^2} \right) < R \right\} \tag{2.84}$$

This result can be extended to an arbitrary number of transmit antennas M_T as follows.

When transmission is constrained to be i.i.d., the outage probability of a slow fading MISO channel for a transmission rate R is:

$$p_{\text{out}}(R) = \Pr \left\{ \log_2 \left(1 + \frac{\bar{P}}{M_T} \frac{\|\mathbf{h}\|^2}{\sigma_n^2} \right) < R \right\} \tag{2.85}$$

Optimal space time codes achieving the outage probability for a number of transmit antenna larger than 2 are difficult to design. Suboptimal designs are described in Chapter 3.

We close this section with the following remarks.

▶ Compared to the case when the CSIT is known at the transmitter, there is a performance loss of $1/M_T$ in the SNR.
▶ Optimal codes extract a full diversity gain of M_T.
▶ When the channel has an i.i.d. Rayleigh fading distribution, an i.i.d. transmission with correlation matrix $\mathbf{R}_{xx} = \bar{P}/M_T \mathbf{I}$ is not optimal in general. However, it becomes optimal asymptotically at high SNR, so that the outage probability can be approximated as (2.85).

2.7.6   Slow Fading MIMO Channel with CSIT Distribution

A slow fading MIMO channel benefits from both diversity and multiplexing gain in order to minimise the outage probability. The optimal transmitter is D-Blast (described in Chapter 3). D-Blast extracts the maximum diversity gain of $M_T M_R$, for an i.i.d Rayleigh fading channel. Although it still benefits from a full multiplexing gain, it has to trade some throughput for diversity gain to improve communication reliability. On the opposite side, V-Blast, which is optimal for a fast fading MIMO channel, does not need to exploit space diversity because a reliable communication is possible by coding over many channel fades. Hence it targets a high throughput.

2.7.6.1   Outage Probability for a General MIMO Channel Distribution

When only the CSIT distribution is known, the covariance of the input signals should be adapted to the channel distribution for optimal performance. Let us assume that the covariance matrix of the input signal is fixed. For a fixed input covariance matrix, the maximal achievable rate for a reliable communication over a given channel fade H is equal to $\mathcal{R}_{\max} = \log_2 \det[I + \frac{1}{\sigma_n^2} H R_{xx} H^H]$. Assume that a rate R is selected at the transmitter according to the channel distribution. Over a channel fade, the system is in outage when the selected transmission rate is larger than $\mathcal{R}_{\max}$. Hence, the outage probability for transmission rate R and a given correlation matrix R_{xx} is:

$$p_{\text{out}}(R, R_{xx}) = \Pr \left\{ \log_2 \det \left[I + \frac{1}{\sigma_n^2} H R_{xx} H^H \right] < R \right\} \tag{2.86}$$

Now, to minimise outage probability, the covariance matrix R_{xx} should also be optimised.

The outage probability of a slow fading MIMO channels with CSIT distribution for a transmission rate R is defined as:

$$p_{\text{out}}(R) = \min_{R_{xx} \text{ s.t. } \text{tr}(R_{xx}) \leq \bar{P}} \Pr \left\{ \log_2 \det \left[I + \frac{1}{\sigma_n^2} H R_{xx} H^H \right] < R \right\} \tag{2.87}$$

For a fixed input covariance matrix, the capacity with outage p_o is the value $C_{\text{out}}(p_o, R_{xx})$ such that:

$$\Pr \left\{ \log_2 \det \left[I + \frac{1}{\sigma_n^2} H R_{xx} H^H \right] < C_{\text{out}}(p_o, R_{xx}) \right\} = p_o \tag{2.88}$$

Hence, the capacity with outage is:

$$\max_{R_{xx} \text{ s.t. } \mathrm{tr}(R_{xx}) \leq \bar{P}} C_{\mathrm{out}}(p_o, R_{xx}) \tag{2.89}$$

The optimal correlation matrix depends on the statistics of the channel and is not easy to find in general for an arbitrary channel distribution. However, when the transmission is constrained to be i.i.d, the computation of the outage probability becomes simplified.

2.7.6.2   Outage Probability for an i.i.d Transmission

When the transmission is constrained to be i.i.d ($R_{xx} = \bar{P}/M_{\mathrm{T}}I$), the outage probability of a slow fading MIMO channel with CSIT distribution for transmission rate R is:

$$p_{\mathrm{out}}(\mathrm{R}) = \Pr \left\{ \log_2 \det \left[1 + \frac{\bar{P}}{M_{\mathrm{T}}\sigma_n^2} HH^{\mathrm{H}} \right] < \mathrm{R} \right\} \tag{2.90}$$

2.7.6.3   Outage Probability for Rayleigh i.i.d MIMO Channel

An i.i.d transmission is optimal for a fast fading Rayleigh channel. It is not for a slow fading channel in general. For MISO channels, $R_{xx} = \bar{P}/M_{\mathrm{T}}I$ is the optimal correlation matrix at high SNR. When the receiver possesses multiple antennas, this is not true anymore. However, (2.90) is considered as a good upper bound on the outage probability at high SNR.

2.7.6.4   Performance: Outage Probability for a Rayleigh i.i.d Slow Fading MIMO Channel

A common way to illustrate the capacity benefits of MIMO slow fading systems is through the use of CDFs. Figure 2.19 shows the CDF of the maximal achievable gain for: a) a SISO channel, b) a 2×2 MIMO channel and c) a 4×4 MIMO channel. All the channels have Rayleigh statistics with $E\left[|h_{ij}|^2\right] = 1$ and the average SNR per link, $\rho = \bar{P}/\sigma_n^2$, is equal to 10 dB.

Looking at the CDF curves, the corresponding outage probability for a fixed selected rate can be easily found. We can observe that the outage probability decreases quite dramatically as the MIMO dimensions increase. For example, for a selected rate R = 5 bits/transmission, the outage probability is 0.95 for a SISO system, 0.35 for a 2×2 MIMO system, while it is much smaller for a 4×4 MIMO system.

The capacity with outage can also be determined as shown in the figure where a fixed outage probability of 10% is considered. The capacity with 10% outage is equal to 1 bit per transmission for the SISO system, 3.9 bits per transmission for the 2×2 MIMO system and 9.3 bits per transmission for the 4×4 MIMO system.

Figure 2.20 shows the capacity with 10% outage as a function of the SNR ρ. With this level of outage, the curves look very similar to the fast fading curve in Figure 2.16. The

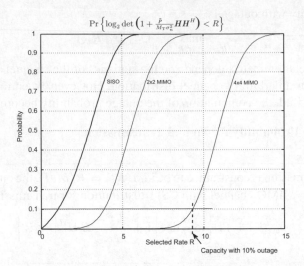

Figure 2.19 Cumulative distribution functions of the maximal achievable rate at 10 dB for a SISO channel, a 2×2 MIMO channel and a 4×4 MIMO channel. The capacity with 10% outage can easily be determined from the CDF.

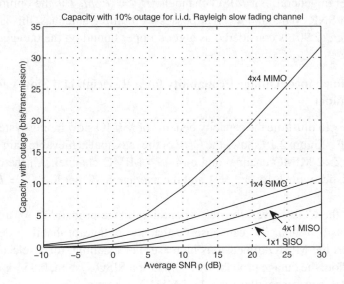

Figure 2.20 Capacity with outage at 10 dB.

capacity with outage does not offer as much information as the outage probability. The performance measure of choice is actually the outage probability as shown in Figure 2.21, where we can observe that, at high SNR, the slope of the probability curves is equal to the diversity gain.

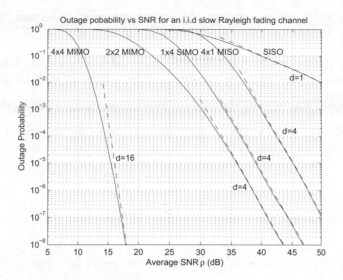

Figure 2.21 Outage probability: the diversity gain can be determined from the curve slopes.

2.7.6.5 Mathematical Definition of Diversity Gain

In many systems, the diversity gain can be easily determined by examining the number of independent channels carrying each codeword. However, for more complex systems, the following mathematical definition becomes easier to manipulate. Diversity gain is defined at high SNR only, although the notion of diversity is meaningful at low SNR as well.

> The diversity gain (DG) of a slow fading channel MIMO channel with outage probability $p_{out}(\rho)$ is:
>
> $$\lim_{\rho \to \infty} \frac{\log_2 p_{out}(\rho)}{\log_2 \rho} = -DG \tag{2.91}$$

The outage probability depends on the selected transmission rate, however, the diversity gain does not.

Summary for slow fading channels with CSIR, distribution of CSIT

For slow fading channels, coding spreads over a block of data where the channel remains constant. As the CSIT is not available, a reliable communication is not possible. The performance measures for slow fading channels are:

▷ The *outage probability*. For a selected transmit rate R, it is the probability that R is larger than the maximal achievable rate for reliable communication.

▶ The *capacity with outage*. A targeted outage probability p_o is set by the system. The capacity with outage p_o is the maximal rate for which is the outage probability is smaller than p_o.

Both quantities can be determined at the transmitter using the distribution of the CSIT allowing the transmitter to select its transmission rate matching a targeted outage probability.

▶ SISO/SIMO channel $\mathbf{y}(k) = \mathbf{h}\,x(k) + \mathbf{n}(k)$.
  The outage capacity of a slow fading SISO/SIMO channel for target rate R is:

$$p_{\text{out}}(\text{R}) = \Pr\left\{\log_2\left(1 + \frac{\bar{P}\|\mathbf{h}\|^2}{\sigma_n^2}\right) < \text{R}\right\} \tag{2.92}$$

In the capacity achieving strategy, the transmit power is kept constant over time and equal to its maximum $\bar{P}$. The code minimising the outage probability is the AWGN code.

▶ MISO channel $y(k) = \mathbf{h}^T\mathbf{x}(k) + n(k)$.
  For an i.i.d transmission (i.e. $\mathbf{R}_{xx} = \bar{P}/M_T I$), the outage probability is:

$$p_{\text{out}}(\text{R}) = \Pr\left\{\log_2\left(1 + \frac{\bar{P}}{M_T}\frac{\|\mathbf{h}\|^2}{\sigma_n^2}\right) < \text{R}\right\}. \tag{2.93}$$

For a 2×1 system, the Alamouti space–time code minimises the outage probability.

▶ MIMO channel $\mathbf{y}(k) = \mathbf{H}\mathbf{x}(k) + \mathbf{n}(k)$.
  For a general MIMO channel distribution, the outage probability is:

$$p_{\text{out}}(\text{R}) = \min_{\mathbf{R}_{xx}\ \text{s.t. } \text{tr}(\mathbf{R}_{xx})\leq\bar{P}} \Pr\left\{\log_2\det\left[\mathbf{I} + \frac{1}{\sigma_n^2}\mathbf{H}\mathbf{R}_{xx}\mathbf{H}^{\text{H}}\right] < \text{R}\right\} \tag{2.94}$$

For an i.i.d transmission, the outage probability is:

$$p_{\text{out}}(\text{R}) = \Pr\left\{\log_2\det\left[\mathbf{I} + \frac{\bar{P}}{M_T\sigma_n^2}\mathbf{H}\mathbf{H}^{\text{H}}\right] < \text{R}\right\} \tag{2.95}$$

For a i.i.d Rayleigh fading channel, (2.95) is a good upper bound on the outage probability at high SNR.
  D-Blast (described in Chapter 3) is a MIMO optimal transceiver architecture.

2.8   Chapter Summary Tables

The main results on capacity contained in this chapter are presented in the form of tables for easy reference, according to three types of systems, three types of channel fading and two types of CSIT as described below.

▶ Three types of systems:

1. **SISO/SIMO**: $\mathbf{y}(k) = \mathbf{h}(k)x(k) + \mathbf{n}(k)$.
    SISO and SIMO are treated together because the SIMO channel can be made equivalent to a SISO channel when the CSI is known at the receiver (by performing spatial matched filtering which is information lossless).
2. **MISO**: $y(k) = \mathbf{h}(k)^T \mathbf{x}(k) + n(k)$.
3. **MIMO**: $\mathbf{y}(k) = H(k)\mathbf{x}(k) + \mathbf{n}(k)$.

For a fading channel, the channel matrix is a random variable denoted as $\mathbf{h}(k)$ (MISO/SIMO) and $H(k)$ (MIMO), while the channel realisation is denoted as $h(k)$ and $H(k)$.

▶ Three types of channel fading:

1. **Time-invariant channel**.
2. **Fast fading channel**: code spans over many channel coherence periods.
3. **Slow fading channel**: code spans over a single channel coherence period.

▶ Two types of CSIT:

1. **Instantaneous value of CSIT**.
2. **Distribution of CSIT**.

In general, the CSIT groups the channel coefficients as well as the noise variance σ_n^2. The CSIR (channel coefficients) is assumed perfectly known at the receiver.

▶ Important quantities:
   ▷ Singular value decomposition: $H = U\Lambda V^H$, eigenvalues are denoted as λ_k.
   ▷ SNR per eigenchannel: $\gamma_k = \lambda_k/\sigma_n^2$.
   ▷ Rank of channel matrix H: r_H, which determines the multiplexing gain for a time-invariant channel.
   ▷ $M_{\min} = \min(M_T, M_R)$ determines the multiplexing gain for a fading channel. M_T and M_R are the number of transmit and receive antennas.
   ▷ $\rho = \bar{P}/\sigma_n^2$, where $\bar{P}$ is the maximal transmit power.

General results on SISO/SIMO capacity

Performance measure		
	Distribution of CSIT, CSIR	CSIT and CSIR
Time invariant (AWGN)		**Shannon capacity** $C_{SIMO}^{TI} = \log_2 \left(1 + \dfrac{\bar{P}\|\mathbf{h}\|^2}{\sigma_n^2} \right)$
Fast fading	**Ergodic capacity** $C_{SIMO}^{FF} = E\left[\log_2 \left(1 + \dfrac{\bar{P}\|\mathbf{h}\|^2}{\sigma_n^2} \right) \right]$	
Slow fading	**Outage probability for rate** R $p_{out}(R) = \Pr\left\{ \log_2 \left(1 + \dfrac{\bar{P}\|\mathbf{h}\|^2}{\sigma_n^2} \right) < R \right\}$ **Capacity with outage** p_o $C_{SIMO}^{SF}(p_o)$ verifying $p_{out}(C_{SIMO}^{SF}(p_o)) = p_o$	

Properties of capacity achieving transceiver		
	Distribution of CSIT, CSIR *Rate optimally adjusted according to CSIT distribution*	CSIT and CSIR *Rate (and power) optimally adjusted based on instantaneous value of CSIT*
Time invariant (AWGN)		Receiver: **spatial matched filter** **AWGN code**: input stream with fixed rate equal to capacity, fixed power equal to $\bar{P}$, coding across noise
Fast fading	Receiver: **spatial matched filter** Input stream with fixed rate equal to capacity, fixed power equal to $\bar{P}$, coding across noise and channel fades	
Slow fading	same as TI channel	

General results on MISO capacity

Performance measure		
	Distribution of CSIT, CSIR	CSIT and CSIR
Time invariant		**capacity** $$C_{\text{SIMO}}^{\text{TI}} = \log_2\left(1 + \frac{\bar{P}\|\mathbf{h}\|^2}{\sigma_n^2}\right)$$
Fast fading	**Ergodic capacity** $$C_{\text{SIMO}}^{\text{FF}} = E\left[\log_2\left(1 + \frac{1}{M_{\text{T}}}\frac{\bar{P}\|\mathbf{h}\|^2}{\sigma_n^2}\right)\right]$$	
Slow fading	**Outage probability for rate R** $$p_{\text{out}}(\text{R}) = \Pr\left\{\log_2\left(1 + \frac{1}{M_{\text{T}}}\frac{\bar{P}\|\mathbf{h}\|^2}{\sigma_n^2}\right) < \text{R}\right\}$$ **Capacity with outage p_{o}** $$C_{\text{SIMO}}^{\text{SF}}(p_{\text{o}}) \text{ verifying } p_{\text{out}}(C_{\text{SIMO}}^{\text{SF}}(p_{\text{o}})) = p_{\text{o}}$$	

Properties of capacity achieving transceiver		
	Distribution of CSIT, CSIR *Rate and input covariance matrix optimally adjusted according to CSIT distribution*	CSIT and CSIR *Rate, power and input covariance matrix optimally adjusted based on instantaneous value of CSIT*
Time invariant		Transmitter: **spatial matched filter** Input stream: AWGN code
Fast fading	Special case of V-Blast with 1 receive antenna (see Chapter 3)	
Slow fading	For 2×1 MISO and i.i.d. transmission: Alamouti STBC (see Chapter 3)	

General results on MIMO capacity

Performance measure

	Distribution of CSIT, CSIR	CSIT and CSIR
Time invariant		**Capacity** $$C_{MIMO}^{TI} = \sum_{k=1}^{r_H} \log_2\left(1 + P_k^o \gamma_k\right)$$ $$P_k^o = \left(\frac{1}{\gamma_0} - \frac{1}{\gamma_k}\right)^+$$ $\gamma_k = \lambda_k/\sigma_n^2$ and γ_0 verifies $\sum_{k=1}^{r_H} P_k^o = \bar{P}$
Fast fading	**Ergodic Capacity** General channel distribution: $$C_{MIMO}^{FF} = \max_{R_{xx},\,\mathrm{tr}(R_{xx})\le\bar{P}} E\left[\log_2\det\left[I + \frac{1}{\sigma_n^2}HR_{xx}H^H\right]\right]$$ i.i.d. Rayleigh Fading: $$C_{MIMO}^{FF} = E\left[\log_2\det\left[I + \frac{\bar{P}}{M_T\sigma_n^2}HH^H\right]\right]$$	
Slow fading	**Outage probability for rate R** General channel distribution: $$\min_{R_{xx},\,\mathrm{tr}(R_{xx})\le\bar{P}} \Pr\left\{\log_2\det\left[I + \frac{1}{\sigma_n^2}HR_{xx}H^H\right] < R\right\}$$ Isotropic transmission: $$\Pr\left\{\log_2\det\left[I + \frac{\bar{P}}{M_T\sigma_n^2}HH^H\right] < R\right\}$$	

Properties of capacity achieving transceiver

	Distribution of CSIT, CSIR Rate per stream and input covariance matrix optimally adjusted according to CSIT distribution	CSIT and CSIR Rate and power per stream, input covariance matrix optimally adjusted according to instantaneous CSIT
Time invariant		Eigenmode transmission: r_H independent AWGN coded input streams
Fast fading	**V-Blast** (see Chapter 3) Independent AWGN coded input streams. SIC at receiver	
Slow fading	**D-Blast** (see Chapter 3) Diagonal encoding and decoding	

> ## MIMO performance gains

Performance of MIMO systems are assessed by three types of gains. The gains are defined when comparing to a single link communication.

▶ **Multiplexing gain.** The multiplexing gain is the number of independent streams that can be transmitted simultaneously. The multiplexing gain of a MIMO system with capacity $C(\rho)$ is defined as:

$$\lim_{\rho \to \infty} \frac{C(\rho)}{\log_2 \rho} \tag{2.96}$$

where $\rho = \bar{P}/\sigma_n^2$. For a time-invariant channel, the maximal multiplexing gain is equal to the rank of the channel matrix. For a fading channel, the maximal multiplexing gain is equal to $\min(M_T, M_R)$, provided a rich scattering environment.

▶ **Array gain.** The array gain is usually defined for a MIMO system transmitting a single stream (using beamforming techniques as described in Chapter 3). It is defined as the increase of the post-processing SNR obtained from using multiple antennas at the transmitter or receiver side. This gain can be extracted when the CSI is known by an appropriate combining of the transmitted or received signals. The maximal array gain is λ_{max}^2 where λ_{max} is the maximal singular value of the channel matrix: it is extracted when the CSI is known at both the transmitter and the receiver using maximal eigenmode transmission (described in Chapter 3).

For a multiple stream MIMO communication, it is uneasy to define an array gain compared to a single link communication, because multiple post-processing SNRs associated to each stream exist. An array gain per stream could be defined that is useful when comparing systems with the same number of multiplexed streams.

▶ **Diversity gain.** The diversity gain is defined for slow fading channels with no CSIT where diversity is used to increase the communication reliability (decrease of the outage probability). It is defined as the number of independent channels carrying the same information. The maximal diversity gain is equal to $M_R M_T$. The diversity gain d of a slow fading channel MIMO channel with outage probability $p_{out}(\rho)$ is defined mathematically at high SNR as:

$$\lim_{\rho \to \infty} \frac{\log_2 p_{out}(\rho)}{\log_2 \rho} = -\text{DG} \tag{2.97}$$

Asymptotic MIMO performance gains

Time-invariant channel

	Low SNR	High SNR
Asymptotic value of capacity	$C \sim \lambda_{max}^2 \rho / \log 2$ Dominant gain = tx and and rx array gain	$C \sim r_H \log_2 \rho$ Dominant gain = multiplexing C scales linearly with r_H
SM gain	None	Channel rank r_H
Array gain	λ_{max}^2	Negligible w.r.t. to multiplexing gain
Channel conditioning	No requirement	Well conditioned (for high performance)

Fast fading channel

	Low SNR	High SNR
Asymptotic value of capacity	$C \sim M_R \rho / \log 2$ Dominant gain = rx array gain Equivalent to SIMO	$C \sim \min(M_T, M_R) \log_2 \rho$ Dominant gain = multiplexing C scales linearly with $\min(M_T, M_R)$
SM gain	None	$\min(M_T, M_R)$
Array gain	M_R Negligible gain when M_T increases	Negligible w.r.t. to multiplexing gain
Channel conditioning	No requirement	Statistically well conditioned (for high performance)

Slow fading channel

	Low SNR	High SNR
Outage probability	Dominant gains = array and diversity gains	Dominant gains = multiplexing and diversity gain
SM gain	None	M_R
Array gain	M_R	Negligible w.r.t. to multiplexing gain
Div. gain	M_R	$M_R M_T$
Channel conditioning	No requirement	Statistically well conditioned (for high performance)

2.9   Further Reading

The capacity of a fixed channel with additive white Gaussian noise was first derived by Shannon in his seminal paper. It has been known to be the upper limit of the achievable communication rate over a fixed channel and has been used as a a benchmark for the performance of various coding and modulation schemes (Shannon 1948). However, as wireless communications came into place and it became apparent that the channel was not a constant quantity, the effects of a time-varying, that is, fading, channel became the object of theoretical and practical study (Goldsmith (1997), Goldsmith and Chua (1997, 1998), Goldsmith and Varaiya (1997)).

The topic of multiple antenna communications emerged in the mid 1990s as a promising technique to increase the capacity (Foschini and Gans (1998), Telatar and Tse (2000)), although multiple antenna techniques in the context of diversity systems had been known for years. The initial theoretical results motivated a series of research projects that focused on different aspects of initially single link communications, that is, D-Blast and V-Blast (Foschini (1996), Foschini et al. (1998), Golden et al. (1999), Wolniansky et al. (1998)).

The idea of eigenmode transmission was investigated for example, in Bach Andersen 2000a,b, while the idea of applying waterfilling techniques (taking into account the cross-correlation of the interference) was investigated in Chizhik, Farrokhi et al. (2000) and Farrokhi et al. (2001).

The topic of transmit antenna diversity was studied, among others, in Tarrokh et al. (1999), while the trade-off between diversity and multiplexing was analysed in Tse et al. (2004).

3

MIMO Transceivers

Elisabeth De Carvalho

In the previous chapter, we have presented performance aspects of MIMO communications based on information theoretic tools, such as capacity or outage probability. Those tools are simple and yet very powerful as they allow for a performance characterisation without the need to design the transmitted signals and the receiver. However, information theoretic tools are also somewhat limited as they make abstraction of the constraints involved in the design of a practical system. This chapter takes a closer look at the transmitter and receiver design. The primary focus is on transceiver structures that allow for data multiplexing. A secondary focus is on MIMO diversity techniques improving communication reliability in a slow fading environment.

1. **Receivers.**
   We first examine the receiver side and describe some of the most practical MIMO receivers:
   - Linear receivers: each of the transmitted multiplexed data streams are estimated at the receiver as a linear combination of the received signals. We describe the matched filter, the zero-forcing and the minimum mean square error (MMSE) linear receivers.
   - Receivers based on successive interference cancellation: in particular, we describe V-Blast, an optimal transceiver for fast fading channels aiming at high throughput.
2. **Joint transmitter and receiver design with CSI.** Assuming the knowledge of the CSI at both ends of the communication link, transmission and reception can be adapted to the channel. Those techniques are called beamforming in this chapter. We describe several beamforming schemes. In particular, we describe the eigenmode transmission scheme which is a MIMO multiplexing architecture based on multiple transmit and receive beams where each beam at the transmitter is matched to a beam at the receiver.

Practical Guide to the MIMO Radio Channel with MATLAB® Examples, First Edition.
Tim Brown, Elisabeth De Carvalho and Persefoni Kyritsi.
© 2012 John Wiley & Sons, Ltd. Published 2012 by John Wiley & Sons, Ltd.

3. **Space–time architectures for slow fading channels.** Coding across time but also across antennas improves the communication reliability, which is needed in a slow fading environment. We describe simple space–time codes called *space–time block codes*. Those codes are designed for MISO systems and do not exploit multiplexing capabilities. At last, we present a space–time architecture called D-Blast where coding is done across time and space and which exploits multiplexing: it provides high throughput but also targets a high diversity gain.

3.1 MIMO Receivers

3.1.1 General MIMO Architecture

To describe the structure of MIMO receivers, we consider a generic MIMO system depicted in Figure 3.1. The M_T transmitted packets (or codewords) are denoted as $X_i, i = 0, \ldots, M_T$ and also referred to as *streams of data*. They are encoded independently. The packets are subjected to the adverse effects of the propagation channel and the additive noise at the receiver. The received packets are denoted as $Y_j, j = 0, \ldots, M_R$. The role of the MIMO receiver is to recover the M_T transmitted packets from the M_R received packets.

During the whole chapter, the processing at the receiver is mostly described on a symbol by symbol basis. A symbol of packet X_i is generically denoted as x_i. Likewise a symbol from the received packet Y_j is denoted as y_j and the corresponding additive noise sample is n_j. Each stream of data is allocated to same power P_x:

$$P_x = \bar{P}/M_T \tag{3.1}$$

where $\bar{P}$ is the total transmit power over all transmit antennas. Corresponding to the transmission and reception via multiple antennas, we define the $M_T \times 1$ complex valued vector of input symbols as **x**, the $M_R \times 1$ complex valued vector of received signals as **y**

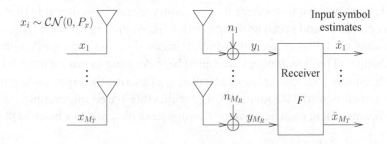

Figure 3.1   General spatial multiplexing structure. M_T packets are sent from M_T transmitting antennas and received at M_R antennas.

and the $M_R \times 1$ complex valued vector of received noise samples as $\mathbf{n}$:

$$\mathbf{x} = \begin{bmatrix} x_1 \\ \vdots \\ x_{M_T} \end{bmatrix}, \qquad \mathbf{y} = \begin{bmatrix} y_1 \\ \vdots \\ y_{M_R} \end{bmatrix}, \qquad \mathbf{n} = \begin{bmatrix} n_1 \\ \vdots \\ n_{M_R} \end{bmatrix} \qquad (3.2)$$

The received signal vector $\mathbf{y}$ contains contributions from all transmitted symbols. The contribution of symbol x_i is $\mathbf{h}_i x_i$ where the $M_R \times 1$ complex valued vector $\mathbf{h}_i$ groups the channel coefficients from transmit antenna i to all receive antennas. The received signal vector can be written as:

$$\mathbf{y} = \sum_{i=1}^{M_T} \mathbf{h}_i x_i + \mathbf{n}, \qquad \mathbf{h}_i = \begin{bmatrix} h_{1i} \\ \vdots \\ h_{M_R i} \end{bmatrix}. \qquad (3.3)$$

Denoting the $M_R \times M_T$ channel matrix as H, Equation (3.3) can be rewritten as a system with vectorial input $\mathbf{x}$ and vectorial output $\mathbf{y}$:

$$\mathbf{y} = H\mathbf{x} + \mathbf{n}, \qquad H = \begin{bmatrix} \mathbf{h}_1 \, \mathbf{h}_2 \cdots \mathbf{h}_{M_T} \end{bmatrix}. \qquad (3.4)$$

The main assumptions in the whole section are summarised as follows:

Main assumptions for MIMO receivers

▶ MIMO channel coefficients are known at the receiver.
▶ Input symbols are independent temporally and spatially: $E\left[\mathbf{x}\mathbf{x}^H\right] = P_x I$.
▶ Noise at receiver is zero mean complex circularly symmetric (ZMCCS), independent temporally and spatially: $\mathbf{n} \sim \mathcal{CN}(0, \sigma_n^2 I)$.

How to extend the theory to an unequal transmit power allocation.
This chapter treats the case of equal transmit power allocation between packets, but only not to overload of the mathematical expressions involved. The results can be easily extended to an unequal power assignment. Assume transmit antenna i is assigned power P_i. For simplicity, we set the symbol power to 1, that is, $E\|x_i\|^2 = P_x = 1$. Hence, the contribution of symbol x_i in the received signal is $\sqrt{P_i}\mathbf{h}_i x_i$. The idea is to integrate the transmit power parameter into the channel. We define the following notations:

$$\mathbf{h}_i' = \sqrt{P_i}\mathbf{h}_i \qquad H' = \begin{bmatrix} \sqrt{P_1}\mathbf{h}_1 \ldots \sqrt{P_{M_T}}\mathbf{h}_{M_T} \end{bmatrix} \qquad (3.5)$$

Then, the input–output relationship is the same as in Equation (3.3): $\mathbf{y} = \sum_{i=1}^{M_T} \mathbf{h}_i' x_i + \mathbf{n} = H'\mathbf{x} + \mathbf{n}$. In the formulas given in this section, H can be replaced by H' and P_x

by 1. Notice that the channel matrix H' can be rewritten as $H' = HP_{\text{root}}$ where P_{root} is a $M_T \times M_T$ diagonal matrix with i^{th} diagonal element equal to $\sqrt{P_i}$.

3.1.2   Maximum Likelihood Receiver

The optimal receiver is the maximum likelihood (ML) receiver. In the most general case, the input symbols belong to a codeword spanning space and time. The number of possible codewords is finite: we denote $\mathcal{CW}$ as the set of all possible codewords. In the ML approach, all possible codewords are tested and the one that best fits the received signal according to the ML criterion is selected as an estimate of the codeword that was actually transmitted.

The codeword transmitted is a matrix denoted as $\mathcal{X}$ with dimension $M_T \times N$. The columns of this matrix are the vectorial inputs of the MIMO system at time instant k to N and are denoted in this section as $\mathbf{x}_k, k = 1, \ldots, N$. The corresponding matrix output is denoted as $\mathcal{Y}$: the columns of $\mathcal{Y}$ are denoted as $\mathbf{y}_k, k = 1, \ldots, N$ where $\mathbf{y}_k$ is the received signal corresponding to the transmission of x_k.

$\mathcal{X}$ is the actual transmitted codeword. When testing all the possible input codewords, a candidate codeword is $\mathcal{Z}$ where $\mathbf{z}_k, k = 1, \ldots, N$ are the columns of $\mathcal{X}$. The ML optimisation criterion finds the codeword with the minimal distance to the received signal:

$$\min_{\mathcal{Z} \in \mathcal{CW}} \|\mathcal{Y} - H\mathcal{Z}\|_F^2 \quad \Leftrightarrow \quad \min_{\{\mathbf{z}_1, \ldots, \mathbf{z}_N\} \in \mathcal{CW}} \sum_{k=1}^{N} \|\mathbf{y}_k - H\mathbf{z}_k\|^2 \qquad (3.6)$$

The ML receiver can be simplified if the receiver proceeds first to estimating the input symbols and then to decoding. Then, ML exploits the fact that a symbol belongs to a constellation (e.g. QPSK) and takes a finite number of values. As symbol x_i takes a finite number of values, so does the transmitted vectorial input $\mathbf{x} = \begin{bmatrix} x_1 \ldots x_{M_T} \end{bmatrix}^T$ (we give up the index k for simplicity). We denote as $\mathcal{C}$ the set of all possible vectorial inputs. Calling $\mathbf{z}$ a candidate vectorial input in the set $\mathcal{C}$, the ML criterion finds the value of $\mathbf{z}$ that best fits the received signal as:

$$\min_{\mathbf{z} \in \mathcal{C}} \|\mathbf{y} - H\mathbf{z}\|^2 \qquad (3.7)$$

The decoder is error free if the value of $\mathbf{z}$ found through the ML optimisation is the actual transmit input $\mathbf{x}$. ML receivers are computationally complex in general because one needs to test all the possible input vectorial values. Their number can become quite large especially as the number of antennas and constellation order increase. Hence, simpler receivers are considered as more practical in general.

3.1.3   Classes of Receivers Considered in the Chapter

In this chapter, we consider the most popular and simple MIMO receivers. Suppose we want to decode packet k. The received signal can be decomposed to isolate the contribution of packet k as:

$$\mathbf{y} = \mathbf{h}_k x_k + \underbrace{\sum_{i \neq k} \mathbf{h}_i x_i}_{\text{Inter-stream interference}} + \mathbf{n} \tag{3.8}$$

The term $\sum_{i \neq k} \mathbf{h}_i x_i$ is an undesired signal and act as interference: we call it inter-symbol interference (ISI).

Suppose the ISI is not present. Then, we have a SIMO channel with input x_k, channel $\mathbf{h}_k$ and additive white Gaussian noise. The optimal receiver is the spatial matched filter (described in Chapter 2), consisting in coherently combining the M_R received signals. Due to the interference, MF is not optimal and leads to poor decoding performance in general. The purpose of the MIMO receivers is to suppress the ISI in order to have estimates of the input symbols with a small estimation error. Following the ISI suppression phase, the input packets are decoded. This section describes two classes of receivers described as follows.

Linear Receivers: A MIMO system can be seen as a linear system of M_R equations (the M_R received signals) with M_T unknowns (the M_T input symbols). When there is no noise, this linear system is simply:

$$\mathbf{y} = \boldsymbol{H}\mathbf{x} \tag{3.9}$$

If there are more equations than unknowns and if the channel has full column rank, this system can be inverted to determine the unknowns: this corresponds to the Zero-Forcing approach. Because of the additive noise, this inversion cannot give perfect estimates of the input symbols. In particular, when the channel matrix is ill-conditioned (the column space of $\boldsymbol{H}$ is close to be rank deficient), estimation performance becomes poor because the channel inversion also enhances the noise. To remedy this problem, the MMSE approach relies on a compromise between system inversion and noise enhancement.

The derivation of linear receivers is based on the theory of *linear estimation*, where the estimates are formed as a linear combination of the observations (the received signals). In our case, each received signal y_j, $j = 1, \ldots, M_R$ contains some information about the transmitted signals. This information is not perfect as the transmitted signals are affected by the channel and corrupted by the additive noise at the receiver. We would like to combine the information from each received signal in the best way possible to extract an estimate of transmitted symbols. A simple way is to perform a linear combination of the received signals. The coefficients of this linear combination are optimised to minimise, in some sense, the error between the estimate of the transmitted symbols and their actual value. They take into account the distortion caused by the MIMO channel but also the noise. Suppose we want to estimate x_k as a linear combination of the received signals y_j,

$j = 1, \ldots, M_R$. The estimate of x_k is of the form:

$$\hat{x}_k = \sum_{j=1}^{M_R} f_{kj} y_j \tag{3.10}$$

The f_{kj}'s are the complex coefficients of the linear combination. We will refer to $\hat{x}_k$ as a *soft estimate* of x_k as opposed to an hard estimate which is the constellation point that is the closest to $\hat{x}_k$.

In this section, the following linear receivers are described:

- Spatial matched filtering (MF) (Section 3.1.4).
- Zero forcing (ZF) (Section 3.1.5).
- MMSE (Section 3.1.6).

Successive Interference Cancellation Receiver: Successive interference cancellation (SIC) is intended for a transmitter structure where each input packet is encoded independently. SIC is an iterative receiver and its principle is as follows: (1) one given packet is decoded using a linear receiver, (2) its contribution in the received signal is reconstructed and removed from the received signal. The next packet is decoded in the same way but with one less interferer. We give the example of V-Blast, a MIMO transceiver architecture whose receiver is based on SIC.

3.1.4   Spatial Matched Filtering

We first describe spatial MF. Although it is intended for a single stream system and not a system with spatial multiplexing, MF is important for two reasons:

1. The linear receivers adapted to spatial multiplexing consists of several processing steps. The first step is the spatial MF.
2. MF is optimal when MIMO channel is orthogonal, i.e the columns of the channel matrix H are orthogonal. This is an important special case as it comprises eigenmode transmission as described in Section 3.2.6.

MF is optimal for a SIMO system provided that the additive noise is white (Section 3.1.4.1). When applied to a MIMO system, spatial MF becomes suboptimal because it does not account for the ISI, treating it as white noise, thus it is optimal only when the MIMO channel is orthogonal.

3.1.4.1   Spatial Matched Filtering for SIMO Systems (with Additive White Noise)

We consider a SIMO system with a single transmit antenna and multiple receive antennas. The received signal at antenna j is $y_j = h_j x + n_j$. We recall that the noise is white:

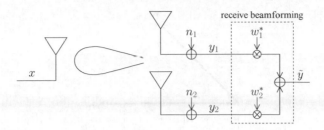

Figure 3.2   Linear receiver for a 1×2 SIMO system. The spatial matched filter maximises the post-processing SNR and corresponds to $w_1^* = h_1^*$ and $w_2^* = h_2^*$.

$\mathbf{n} \sim \mathcal{CN}(0, \sigma_n^2 \mathbf{I})$. Matched filtering consists of a linear combination of the multiple received signals as depicted in Figure 3.2. The received signals y_j are weighted by a complex scalar weight w_j^* and summed up. The resulting signal is: $\sum_{j=1}^{M_R} w_j^* y_j = \left(\sum_{j=1}^{M_R} w_j^* h_j \right) x + \sum_{j=1}^{M_R} w_j^* n_j$. MF is the receiver maximising the post-processing SNR as seen below.

Post-processing SNR: The new system encompassing the spatial filtering at the receiver is a SISO system with channel equal to $\sum_{j=1}^{M_R} w_j^* h_j$. For convenience, we rewrite this equivalent channel as the inner product: $\mathbf{w}^H \mathbf{h}$ where $\mathbf{w}^H = \begin{bmatrix} w_1^* & w_2^* \dots w_{M_R}^* \end{bmatrix}$ is a row vector. The noise part is written as $\mathbf{w}^H \mathbf{n}$. With those notations, the processed received signal has the following expression:

$$\mathbf{w}^H \mathbf{y} = \mathbf{w}^H (\mathbf{h}x + \mathbf{n}) = \mathbf{w}^H \mathbf{h}\, x + \mathbf{w}^H \mathbf{n} \qquad (3.11)$$

The signal power in (3.11) is $P_x |\mathbf{w}^H \mathbf{h}|^2$, while the noise power is $\sigma_n^2 \|\mathbf{w}\|^2$. Hence, the post-processing SNR after spatial filtering is:

$$\text{SNR} = \frac{P_x |\mathbf{w}^H \mathbf{h}|^2}{\sigma_n^2 \|\mathbf{w}\|^2} \qquad (3.12)$$

3.1.4.2   Geometric Interpretation of Maximisation of SNR

The post-processing SNR remains unchanged whatever the value of $\|\mathbf{w}\|^2$. Therefore, we can impose the constraint $\|\mathbf{w}\|^2 = 1$ such that the result of the SNR maximisation remains the same. With this constraint, maximising the post-processing SNR is equivalent to maximising the norm of the inner product $|\mathbf{w}^H \mathbf{h}|$.

Figure 3.3 illustrates a simple case that can be drawn in the 2D plan: two transmit antennas with real channels. The channel vector $\mathbf{h}$ and the weights of the spatial filtering $\mathbf{w}$ are drawn as 2D vectors. From this representation, it can easily be seen that the amplitude of the scalar product $\mathbf{w}^H \mathbf{h}$ is maximised when $\mathbf{h}$ and $\mathbf{w}$ are aligned. The optimal (normalised) weight vector is $\mathbf{w}_{\text{MF}}^H = e^{-j\phi} \mathbf{h}^H / \|\mathbf{h}\|$. As the phase factor does not impact the post-processing SNR, we set the phase ϕ to zero. This result can be extended to the general case.

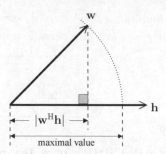

Figure 3.3   SIMO case with two receive antennas and a real channel. The amplitude of the inner product $\mathbf{w}^H\mathbf{h}$ (with $\|\mathbf{w}\|^2 = 1$) is maximised when $\mathbf{h}$ and $\mathbf{w}$ are aligned.

The received filter after the matched filter $\mathbf{w}_{\mathrm{MF}}^H = \mathbf{h}^H/\|\mathbf{h}\|$ is:

$$\tilde{y}_{\mathrm{MF}} = \|\mathbf{h}\|x + \frac{\mathbf{h}^H}{\|\mathbf{h}\|}\mathbf{n} \qquad (3.13)$$

To get an estimate of the transmitted symbol x, an additional scaling is needed to obtain:

$$\hat{x}_{\mathrm{MF}} = x + \frac{\mathbf{h}^H}{\|\mathbf{h}\|^2}\mathbf{n} \qquad (3.14)$$

This last equation describes a channel with an SNR defined as the ratio between the signal power and the noise power and is equal to $P_x\|\mathbf{h}\|^2/\sigma_n^2$. The SNR can equivalently be defined in terms of estimation quantities as the ratio between the signal power and the mean squared estimation error as:

$$\mathrm{SNR}_{\mathrm{MF}} = \frac{P_x}{E|x - \hat{x}_{\mathrm{MF}}|^2} = \frac{P_x\|\mathbf{h}\|^2}{\sigma_n^2} \qquad (3.15)$$

We next summarise the main results about receive matched filtering.

The spatial filter matched to the SIMO channel $\mathbf{h}$ is $\mathbf{w}_{\mathrm{MF}}^H = \mathbf{h}^H/\|\mathbf{h}\|$. The spatial MF maximises the post-processing SNR. The soft estimate of x obtained by the spatial MF is:

$$\hat{x}_{\mathrm{MF}} = x + \frac{\mathbf{h}^H}{\|\mathbf{h}\|^2}\mathbf{n} \qquad \mathrm{SNR}_{\mathrm{MF}} = \frac{P_x\|\mathbf{h}\|^2}{\sigma_n^2} \qquad (3.16)$$

Matched filtering combines the received signals as follows:

1. *Phase alignment*: The phase of the weights w_j^* is adjusted so that the signals $w_j^*h_j$ add constructively. The phase of the signals is set to zero.

2. *Power distribution*: The amplitude of the weights plays also a role in maximising the post-processing SNR. The optimal allocation assigns more power to the stronger channels.

Some remarks:

▶ Spatial MF is information lossless and achieves the SIMO capacity.
▶ The matched filter is frequently defined up to a scale factor as a scaling operation does not change the post-processing SNR. Previously, we have introduced the normalised version of the matched filter. In the rest of the section, the receive matched filter will often be defined as the conjugate transpose of the channel $\mathbf{h}^H$. Then, the received signal after the matched filter $\mathbf{h}^H$ is $\|\mathbf{h}\|^2 x + \mathbf{h}^H \mathbf{n}$. After matched filtering, the signal is sent to a decoder. It is common for decoders to require an entry of the form $x + n'$, where n' is a noise term independent from x. Hence, a proper scaling by $1/\|\mathbf{h}\|^2$ is necessary at the output of the matched filter. Notice that we have applied such a scaling at the output at the normalised matched filter in Equation (3.13) to get a soft estimate of the input symbol.
▶ Spatial matched filtering is also commonly called receive Maximal Ratio Combining (MRC).

3.1.4.3 SIMO Systems with Coloured Noise: $\mathbf{n} \sim \mathcal{CN}(\mathbf{0}, R_{nn})$

We assume now that the noise is a coloured ZMCSC Gaussian random variable. We qualify a random variable as being coloured if it is not white, that is, if its correlation matrix is not a multiple of the identity matrix. This is an important practical case corresponding to a system with interference where the interference is modelled as a coloured Gaussian random variable (see e.g. the MMSE receiver in Section 3.1.6).

A simple matched filtering is suboptimal because there is information in the structure of the noise given by its correlation matrix. A matched filter treats the noise as white and ignores this information. The optimal processing is actually quite intuitive. It consists first of whitening the noise. Once the noise is whitened, the spatial MF as seen previously is applied. Denoting the correlation matrix of the noise as R_{nn}, those two operations are described as follows.

1. *Noise whitening.* The noise is whitened by multiplying the received signal by $R_{nn}^{-1/2}$:

$$\tilde{\mathbf{y}} = R_{nn}^{-1/2} \mathbf{y} = R_{nn}^{-1/2} \mathbf{h}\, x + R_{nn}^{-1/2} \mathbf{n} \qquad (3.17)$$

$R_{nn}^{-1/2}$ comes from the following factorisation of the noise covariance matrix:

$$R_{nn} = R_{nn}^{1/2} R_{nn}^{H/2} \qquad (3.18)$$

One way to obtain such a factorisation is by using the Cholesky decomposition. One can verify that the correlation of the random variable $R_{nn}^{-1/2}\mathbf{n}$ is equal to the identity matrix.

2. *Matched filtering.* The new system in Equation (3.17) corresponds to a SIMO system with channel $R_{nn}^{-1/2}\mathbf{h}$ and white noise for which we know the optimal receiver: it is the filter matched to the channel $R_{nn}^{-1/2}\mathbf{h}$, that is, $\mathbf{h}^H R_{nn}^{-H/2}$.

Noting that $R_{nn}^{-H/2} R_{nn}^{-1/2} = R_{nn}^{-1}$, the signal after noise whitening and matched filtering is:

$$y_{\text{WMF}} = \mathbf{h}^H R_{nn}^{-1}\mathbf{h}\,x + \mathbf{h}^H R_{nn}^{-1}\,\mathbf{n} \qquad (3.19)$$

The spatial whitening MF (WMF) of a SIMO channel with coloured noise $\mathbf{n} \sim \mathcal{CN}(0, R_{nn})$ and corresponding post-processing SNR are:

$$\mathbf{w}_{\text{WMF}}^H = \mathbf{h}^H R_{nn}^{-1} \qquad \text{SNR}_{\text{WMF}} = P_x \mathbf{h}^H R_{nn}^{-1}\mathbf{h} \qquad (3.20)$$

3.1.4.4   Spatial Matched Filtering for MIMO Channels

Orthogonal Channel: Consider the particular case of an orthogonal MIMO channel where the channels of the different streams are orthogonal, that is $\mathbf{h}_i^H\mathbf{h}_{i'} = 0$ for $i \neq i'$. Eigenmode transmission (Section 3.2.6) provides an example of an orthogonal MIMO channel where a pre-processing performed at the transmitter makes the channel orthogonal (seen at the input of the preprocessor). The resulting optimal receiver is very simple: it is the spatial matched filter. Indeed, the matched filter associated with stream k is $\mathbf{h}_k^H$. Applying this matched filtering to the received signal $\mathbf{y}$ eliminates the ISI, because the channels of the interfering streams are orthogonal to $\mathbf{h}_k$. Furthermore, in the absence of inter-stream interference, we know that matched filtering is the optimal receiver.

General Channel: For a nonorthogonal channel, the filter matched to stream k is suboptimal because it does not account for the ISI. The ISI is simply treated as white noise. The output of the matched filter is:

$$\tilde{y}_k = \mathbf{h}_k^H\mathbf{y} = \underbrace{\|\mathbf{h}_k\|^2 x_k}_{\substack{\text{Signal of}\\\text{interest}}} + \underbrace{\sum_{i \neq k}\mathbf{h}_k^H\mathbf{h}_i x_i}_{\substack{\text{Inter-stream}\\\text{interference}}} + \mathbf{h}_k^H\mathbf{n}. \qquad (3.21)$$

The power in the interference term might be significantly high, resulting in poor decoding performance.

Illustrative Example: We illustrate spatial matched filtering for a $2{\times}2$ system with real channels in Figure 3.4. This example will be particularly useful when comparing with the

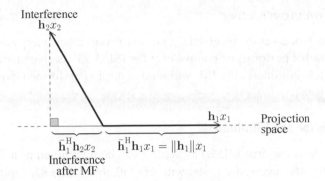

Figure 3.4 Spatial MF for a 2×2 system: to estimate stream 1 with channel $\mathbf{h}_1$, the received signal is projected into the direction of $\mathbf{h}_1$.

ZF and MMSE receivers. We want to perform MF to stream 1 with channel $\mathbf{h}_1$. We define $\bar{\mathbf{h}}_1 = \mathbf{h}_1/\|\mathbf{h}_1\|$, a normalised vector in the direction of $\mathbf{h}_1$. The matched filtering operation can be written as $\bar{\mathbf{h}}_1^H \mathbf{y}$ (we recall that a scaling does not change the post-processing SNR). MF can be seen as the projection of the received signal into the direction of $\mathbf{h}_1$. The interference term $\bar{\mathbf{h}}_1^H \mathbf{h}_2 x_2$ is the component of $\mathbf{h}_2 x_2$ in the direction of $\mathbf{h}_1$.

Spatial matched filtering

▶ In a SIMO system with channel $\mathbf{h}$ and additive white Gaussian noise, the optimal receiver is the matched filter $\mathbf{h}^H$. It maximises the post-processing SNR and achieves capacity.

▶ When the noise is a coloured Gaussian random variable with correlation matrix R_{nn}, the optimal processing is the whitening matched filtering $\mathbf{h}^H R_{nn}^{-1}$ which proceeds to first whitening the noise and then applying the matched filter to the sytem with whitened noise.

▶ For a MIMO system, matched filtering is suboptimal unless the columns of the MIMO channel matrix are orthogonal.

	Spatial MF	Post-processing SNR
SIMO white noise	$\mathbf{h}^H$	$\text{SNR} = \dfrac{P_x}{\sigma_n^2}\|\mathbf{h}\|^2$
SIMO colored noise	$\mathbf{h}^H R_{nn}^{-1}$	$\text{SNR} = P_x \mathbf{h}^H R_{nn}^{-1}\mathbf{h}$
MIMO	$\mathbf{h}_k^H$	$\text{SNR}(k) = \dfrac{P_x\|\mathbf{h}_k\|^2}{\sum_{i \neq k}|\mathbf{h}_k^H\mathbf{h}_i|^2/\|\mathbf{h}_k\|^2 + \sigma_n^2}$

3.1.5 Zero Forcing Receiver

Matched filtering processes the received signals to focus as much energy as possible in the stream of interest, but performs no control over the ISI. A ZF receiver adopts the opposite view. It completely eliminates the ISI, while no control is performed over the energy in the stream of interest.

3.1.5.1 How is the ISI Eliminated?

Let us consider the noise-free MIMO system. It can be interpreted as a linear system of M_R equations (the M_R received signals) with M_T unknowns (the M_T input symbols):

$$\begin{cases} \tilde{y}_1 = h_{11}x_1 + h_{12}x_2 + \cdots + h_{1M_T}x_{M_T} \\ \tilde{y}_2 = h_{21}x_1 + h_{22}x_2 + \cdots + h_{2M_T}x_{M_T} \\ \quad\vdots \\ \tilde{y}_{M_R} = h_{M_R1}x_1 + h_{M_R2}x_2 + \cdots + h_{M_RM_T}x_{M_T} \end{cases} \quad\Longleftrightarrow\quad \tilde{\mathbf{y}} = \mathbf{Hx}. \qquad (3.22)$$

In the ZF approach, this system is inverted to find the unknown input symbols. The ZF receiver is a $M_T \times M_R$ matrix denoted as $\mathbf{F}$, such that:

$$\mathbf{F}\tilde{\mathbf{y}} = \mathbf{x} \quad \text{or} \quad \mathbf{FH} = \mathbf{I} \qquad (3.23)$$

The ZF receiver inverses the MIMO channel matrix. The existence of this channel inversion is subject to some conditions on the channel. Three cases can be distinguished:

1. $\mathbf{H}$ does not have full column rank (e.g. when $M_T > M_R$). There are no unique solutions for the input symbols. A zero-forcing receiver cannot be defined.
2. $\mathbf{H}$ is square and invertible. There exists a single ZF receiver: $\mathbf{F}_{ZF} = \mathbf{H}^{-1}$.
3. $\mathbf{H}$ is tall ($M_T \leq M_R$) and has full column rank. The system is overdetermined. This means that several receivers exist verifying $\mathbf{FH} = \mathbf{I}$. Those receivers can be written as $\left(\mathbf{C}^H \mathbf{H}\right)^{-1} \mathbf{C}^H$, where $\mathbf{C}$ is a $M_T \times M_R$ matrix with full column rank. When there is no additive noise, any of those receivers can be applied to recover the input symbols. However, when there is noise, an error will be made on the estimation of the symbols. The estimation error depends on the coefficients of the receiver. Therefore, an appropriate selection of the ZF receiver is important to get good performance. Next, we describe the ZF receiver that minimises the mean squared estimation error (averaged over the noise) on the symbols.

3.1.5.2 (MMSE-)ZF Receiver

Expression of the Receiver: Let us assume that the channel has full column rank. As just seen, many receivers can eliminate the ISI. A ZF receiver $\mathbf{F}$ verifies $\mathbf{FH} = \mathbf{I}$, hence the output of F is:

$$\hat{\mathbf{x}} = F\mathbf{y} = \mathbf{x} + F\mathbf{n} \tag{3.24}$$

The estimation error is $\hat{\mathbf{x}} - \mathbf{x} = F\mathbf{n}$. The MSE is the expected value of the norm squared of the error, where the expected value is taken with respect to the noise:

$$E_{\mathbf{n}}\|\hat{\mathbf{x}} - \mathbf{x}\|^2 = E_{\mathbf{n}}[\mathbf{n}^{\mathrm{H}} F^{\mathrm{H}} F\mathbf{n}] = \sigma_n^2 \mathrm{tr} F^{\mathrm{H}} F \tag{3.25}$$

This last expression is obtained using the relationship

$$E[\mathbf{n}^{\mathrm{H}} F^{\mathrm{H}} F\mathbf{n}] = E[\mathrm{tr} F\mathbf{n}\mathbf{n}^{\mathrm{H}} F^{\mathrm{H}}] = \mathrm{tr}[F E(\mathbf{n}\mathbf{n}^{\mathrm{H}})F^{\mathrm{H}}] \tag{3.26}$$

The ZF receiver minimising the MSE is unique. We denote it as (MMSE-)ZF receiver or, in the remainder of the chapter, simply as ZF equaliser. It is equal to $F_{\mathrm{ZF}} = (H^{\mathrm{H}} H)^{-1} H^{\mathrm{H}}$. This last expression is the Moore-Penrose pseudo-inverse of H.

SNR per Stream: We now determine the SNR per data stream at the output of the ZF receiver. The kth output of the ZF receiver is of the form:

$$\hat{x}_k = x_k + F_{\mathrm{ZF},k}\mathbf{n} \tag{3.27}$$

where $F_{\mathrm{ZF},k}$ is the kth line of $F_{\mathrm{ZF},k}$. (3.27) defines a SISO channel for which an SNR can be determined. As pointed out in Section 3.1.4, the SNR can also be defined in terms of estimation quantities. In the explanation below, we give up temporarily the ZF subscript as the results are not specific to the ZF receiver and will be applied in the following section treating of the MMSE receiver.

The SNR of stream k is defined as:

$$\mathrm{SNR}(k) = \frac{P_x}{E[|\hat{x}_k - x_k|^2]} \tag{3.28}$$

To obtain a simple expression for $E[|\hat{x}_k - x_k|^2]$, we first compute the $M_{\mathrm{T}} \times M_{\mathrm{T}}$ matrix $\mathcal{E} = E[(\hat{x} - x)(\hat{x} - x)^{\mathrm{H}}]$ and use the relationship $E[|\hat{x}_k - x_k|^2] = \mathcal{E}_{kk}$, that is, $E[|\hat{x}_k - x_k|^2]$ is the kth diagonal element of $\mathcal{E}$.

A simple computation results in the following expression for $\mathcal{E}_{\mathrm{ZF}}$:

$$\mathcal{E}_{\mathrm{ZF}} = \sigma_n^2 (H^{\mathrm{H}} H)^{-1} \tag{3.29}$$

When the channel matrix is full column rank, ZF receivers F can be determined verifying $FH = I$. The (MMSE-)ZF equaliser is the unique ZF receiver minimising the output MSE. Its expression is:

$$F_{\text{ZF}} = \left(H^{\text{H}} H\right)^{-1} H^{\text{H}} \qquad (3.30)$$

The post-processing SNR of the kth output is:

$$\text{SNR}_{\text{ZF}}(k) = \frac{P_x}{\sigma_n^2 \left[K_{\text{ZF}}^{-1}\right]_{kk}} \qquad \text{with } K_{\text{ZF}} = H^{\text{H}} H \qquad (3.31)$$

3.1.5.3 Geometric Interpretation

To give further insights, we provide a geometric interpretation of the ZF operation. While the ZF receiver has been previously presented as a block operation (i.e. operating on the vectorial output), it is now described on a symbol basis. The ZF receiver can be interpreted as a 2-step operation: (1) first it projects the received signal in the direction orthogonal to the interference, thus eliminating the interference, (2) after the projection, the system becomes equivalent to a SIMO system with additive white Gaussian noise for which the optimal processing is matched filtering.

Step 1–Orthogonal Projection to Eliminate the Inter-stream Interference: Assume we want to estimate x_k. Writing the received signal to separate the contribution of x_k and the interference, we get:

$$y = h_k x_k + \sum_{i \neq k} h_i x_i + n \qquad (3.32)$$

The interference $\sum_{i \neq k} h_i x_i$ lies in the subspace spanned by the channel vectors $\{h_i, i \neq k\}$. Let us denote as $\bar{H}_k$ a matrix with $M_T - 1$ columns equal to $\{h_i, i \neq k\}$:

$$\bar{H}_k = \left[h_1 \ldots h_{k-1} h_{k+1} \ldots h_{M_T}\right] \qquad (3.33)$$

The dimension of the interference space is $M_T - 1$. Its orthogonal space has dimension $M_R - M_T + 1$, for which one can find an orthonormal basis, that is, a set of $M_R - M_T + 1$ vectors that have unit norm, are orthogonal and span the entire space. We denote as $\bar{H}_k^{\perp}$ a matrix with columns consists of this orthonormal basis.

As a simple example, let us consider a 2×2 system and the estimation of the first stream with channel h_1. The interference space is spanned by h_2 and has dimension equal to 1. The space orthogonal to h_2 has dimension 1 and is spanned by the vector:

$$\bar{H}_1^{\perp} = h_2^{\perp} = \frac{1}{\|h_2\|} \begin{bmatrix} -h_{21}^* \\ h_{22}^* \end{bmatrix} \qquad (3.34)$$

Two properties of $\bar{H}_k^{\perp}$ are used in what follows:

$$
\begin{cases}
\bar{H}_k^{\perp H} \mathbf{h}_i = 0 & \forall i \neq k \quad \text{(orthogonality)} \\
\bar{H}_k^{\perp H} \bar{H}_k^{\perp} = I & \text{(orthonormality)}
\end{cases}
\tag{3.35}
$$

Multiplying the received signal by $\bar{H}_k^{\perp H}$, we get:

$$
\bar{H}_k^{\perp H} \mathbf{y} = \bar{H}_k^{\perp H} \mathbf{h}_k x_k + \bar{H}_k^{\perp H} \mathbf{n}
\tag{3.36}
$$

Hence, the interference is completely eliminated.

Step 2–Matched Filtering: The new system corresponding to Equation (3.36) is equivalent to a SIMO system with:

- equivalent channel: $\bar{H}_k^{\perp H} \mathbf{h}_k$.
- noise vector: $\bar{H}_k^{\perp H} \mathbf{n}$. As $\bar{H}_k^{\perp H}$ is orthonormal, it correlation matrix is equal to $\sigma_n^2 I$ and $\bar{H}_k^{\perp H} \mathbf{n} \sim \mathcal{CN}(0, \sigma_n^2 I)$.

When $M_T = M_R$, this new system is actually a SISO system, so no additional processing is necessary. Otherwise, we need to further process the multiple outputs in Equation (3.36) to get an estimate of x_k. For such a SIMO system with additive white noise, the optimal receiver is the filter matched to $\bar{H}_k^{\perp H} \mathbf{h}_k$, that is, $\mathbf{h}_k^H \bar{H}_k^{\perp}$ (see Section 3.1.4). The output of the matched filtering is:

$$
\mathbf{h}_k^H \bar{H}_k^{\perp} \bar{H}_k^{\perp H} \mathbf{y} = \mathbf{h}_k^H \bar{H}_k^{\perp} \bar{H}_k^{\perp H} \mathbf{h}_k x_k + \mathbf{h}_k^H \bar{H}_k^{\perp} \bar{H}_k^{\perp H} \mathbf{n}
\tag{3.37}
$$

The product $P_{\bar{H}_k^{\perp}} = \bar{H}_k^{\perp} \bar{H}_k^{\perp H}$ is the orthogonal projection into the columns of $\bar{H}_k^{\perp H}$ (see Appendix (Some Useful Definitions)). $P_{\bar{H}_k^{\perp}}$ can be expressed in terms of the channel matrix directly:

- Firstly, the orthogonal projection into the columns of $\bar{H}_k$ is: $P_{\bar{H}_k} = \bar{H}_k \left(\bar{H}_k^H \bar{H}_k \right)^{-1} \bar{H}_k^H$.
- The orthogonal projection into the space orthogonal to the columns of $\bar{H}_k$ is:

$$
P_{\bar{H}_k}^{\perp} = I - P_{\bar{H}_k} = I - \bar{H}_k \left(\bar{H}_k^H \bar{H}_k \right)^{-1} \bar{H}_k^H
\tag{3.38}
$$

Continuing with the simple example of a 2×2 system, $P_{\bar{H}_1} = P_{\mathbf{h}_2} = \mathbf{h}_2 \mathbf{h}_2^H / \|\mathbf{h}_2\|^2$ and $P_{\bar{H}_1}^{\perp} = P_{\mathbf{h}_2}^{\perp} = I - \mathbf{h}_2 \mathbf{h}_2^H / \|\mathbf{h}_2\|^2$.

Using the relationship $\bar{\boldsymbol{H}}_k^{\perp} \bar{\boldsymbol{H}}_k^{\perp\text{H}} = \boldsymbol{P}_{\bar{H}_k}^{\perp}$, the output of the matched filtering is:

$$\mathbf{h}_k^{\text{H}} \boldsymbol{P}_{\bar{H}_k}^{\perp} \mathbf{y} = \mathbf{h}_k^{\text{H}} \boldsymbol{P}_{\bar{H}_k}^{\perp} \mathbf{h}_k x_k + \mathbf{h}_k^{\text{H}} \boldsymbol{P}_{\bar{H}_k}^{\perp} \mathbf{n} \qquad (3.39)$$

We need to divide by $\mathbf{h}_k^{\text{H}} \boldsymbol{P}_{\bar{H}_k}^{\perp} \mathbf{h}_k$, a soft estimate of the kth stream is:

$$\hat{x}_{\text{ZF},k} = x_k + \frac{\mathbf{h}_k^{\text{H}} \boldsymbol{P}_{\bar{H}_k}^{\perp}}{\mathbf{h}_k^{\text{H}} \boldsymbol{P}_{\bar{H}_k}^{\perp} \mathbf{h}_k} \mathbf{n} \qquad (3.40)$$

The ZF receiver used to estimate the kth stream is:

$$\boldsymbol{F}_{\text{ZF}}(k) = \frac{\mathbf{h}_k^{\text{H}} \boldsymbol{P}_{\bar{H}_k}^{\perp}}{\mathbf{h}_k^{\text{H}} \boldsymbol{P}_{\bar{H}_k}^{\perp} \mathbf{h}_k} \qquad (3.41)$$

The post-processing SNR of the kth stream is:

$$\text{SNR}_{\text{ZF}}(k) = \frac{P_x}{\sigma_n^2} \mathbf{h}_k^{\text{H}} \boldsymbol{P}_{\bar{H}_k}^{\perp} \mathbf{h}_k \qquad (3.42)$$

Expressions (3.41) and (3.42) are equivalent to (3.30) and (3.31): (3.41) is the kth line of $\boldsymbol{F}_{\text{ZF}}$ in (3.30) and $\mathbf{h}_k^{\text{H}} \boldsymbol{P}_{\bar{H}_k}^{\perp} \mathbf{h}_k$ in (3.42) is a closed form expression of the kth diagonal element of $\boldsymbol{K}_{\text{ZF}}^{-1} = \left(\boldsymbol{H}^H \boldsymbol{H}\right)^{-1}$.

Illustrative Example: We illustrate the ZF operation for a 2×2 system with real channels in Figure 3.5. The received signal is $\mathbf{h}_1 x_1 + \mathbf{h}_2 x_2 + \mathbf{n}$. The purpose is to detect stream 1 with channel $\mathbf{h}_1$. Stream 2 with channel $\mathbf{h}_2$ is an interferer. The component of $\mathbf{h}_2$ into the space orthogonal to the interferer is null. Notice that the norm of component of $\mathbf{h}_1$ gets reduced compared to matched filtering (Section 3.1.4).

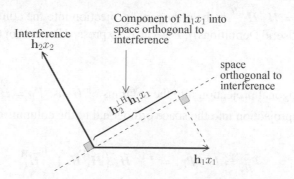

Figure 3.5 ZF operation for a 2×2 system: to estimate stream 1 with channel $\mathbf{h}_1$, the received signal is projected into the space orthogonal to the interference.

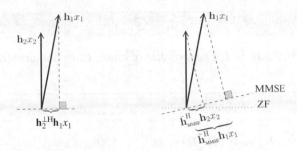

Figure 3.6 ZF receiver and noise enhancement (left): when the direction of $\mathbf{h}_1$ is close to the direction of the interferer, the projection of $\mathbf{h}_1$ into the space orthogonal to the interference is very small. The MMSE receiver (right) offers a compromise between noise enhancement and interference suppression.

3.1.5.4 Noise Enhancement

The major drawback of the ZF receiver is what is called *noise enhancement*. Although the ZF receiver eliminates the interference, performance becomes poor when the channel of the signal of interest is almost colinear to the interference subspace, or in other words when the channel matrix is almost rank deficient, as illustrated in Figure 3.6.

In this figure, the stream of interest is stream 1 while stream 2 interferes. The direction of the stream to decode and the direction of the interference are very close. As a consequence, the component of the stream to decode into the space orthogonal to the interference is small (see figure on the left). After projection, the signal is: $\mathbf{h}_2^{\perp H}\mathbf{h}_1 x_1 + \mathbf{h}_2^{\perp H}\mathbf{n}$. After proper scaling, the received signal is:

$$\tilde{\mathbf{y}}_1 = x_1 + \frac{\mathbf{h}_2^{\perp H}}{\mathbf{h}_2^{\perp H}\mathbf{h}_1}\mathbf{n} \tag{3.43}$$

When the term $\mathbf{h}_2^{\perp H}\mathbf{h}_1$ is very small, the noise term becomes very large and the post-processing SNR very low.

The receiver that is described now, the MMSE receiver, remedies to the noise enhancement problem by offering a compromise between noise enhancement and interference suppression. The MMSE receiver is usually the receiver of choice (especially when coupled with successive interference cancellation). The ZF receiver is, however, useful for analysis purposes: it allows for closed form expressions of useful performance measures.

ZF receiver
The ZF receiver is only defined when the channel matrix $\boldsymbol{H}$ has full column rank. It minimises the output MSE under the constraint of complete elimination of the

ZF receiver (Continued)		

interference, that is, $F_{ZF}H = I$. The formulas of interest are summarised in the following table.

	Receiver	Post-processing SNR
Block-wise	$F_{ZF} = H\left(H^H H\right)^{-1} H^H$	$\mathrm{SNR}_{ZF}(k) = \dfrac{P_x}{\sigma_n^2\left[\left(H^H H\right)^{-1}\right]_{kk}}$
Stream-wise	$F_{ZF}(k) = \dfrac{\mathbf{h}_k^H P_{\bar{H}_k}^\perp}{\mathbf{h}_k^H P_{\bar{H}_k}^\perp \mathbf{h}_k}$	$\mathrm{SNR}_{ZF}(k) = \dfrac{P_x}{\sigma_n^2}\mathbf{h}_k^H P_{\bar{H}_k}^\perp \mathbf{h}_k$

3.1.6 MMSE Receiver

The purpose of the MMSE receiver is to minimise the average estimation error on the transmitted symbols. The average is taken over the transmitted symbols and the noise: the MSE is $E_{\mathbf{x},\mathbf{n}}\|\hat{\mathbf{x}} - \mathbf{x}\|^2$. We recall that the ZF receiver also minimises the output MSE but under the constraint of complete ISI elimination. Thus, it is different from the MMSE receiver.

MMSE Approach: The output of a receiver F is of the form $F\mathbf{y}$ and the vector of errors is

$$\hat{\mathbf{x}} - \mathbf{x} = (FH - I)\mathbf{x} + F\mathbf{n} \tag{3.44}$$

Using the relationship $\mathbf{n}^H F^H F\mathbf{n} = \mathrm{tr}F\mathbf{n}\mathbf{n}^H F^H$, the MMSE cost function (i.e the MSE) is:

$$E_{\mathbf{x},\mathbf{n}}\left[\|\hat{\mathbf{x}} - \mathbf{x}\|^2\right] = P_x(FH - I)^H(FH - I) + \sigma_n^2 \mathrm{tr}F^H F \tag{3.45}$$

Minimising (3.45) w.r.t. F leads to the following expression for the MMSE receiver:

$$F_{MMSE} = P_x H^H R_{yy}^{-1} \qquad \text{where} \quad R_{yy} = E\left[\mathbf{y}\mathbf{y}^H\right] = P_x HH^H + \sigma_n^2 I \tag{3.46}$$

R_{yy} is the covariance matrix of the received signal. Using the matrix inversion lemma (see Appendix (Some Useful Definitions)), an equivalent expression of the MMSE receiver is:

$$F_{MMSE} = \left(H^H H + \frac{\sigma_n^2}{P_x}I\right)^{-1} H^H \tag{3.47}$$

The advantage of this modified expression is the complexity involved in the inversion: when $M_T < M_R$, the dimension of the matrix to be inverted is smaller than the one in equation (3.46). Strictly speaking, the MMSE receiver does not require the channel matrix to have full column rank. However, performance becomes poor when the channel does not have full column rank.

Bias of the MMSE Receiver: Based on the expression (3.46), the kth line of F_{MMSE} is $P_x \mathbf{h}_k^H R_{\text{yy}}^{-1}$. Hence, the kth output of the MMSE receiver is:

$$\hat{x}_{\text{MMSE},k} = \mathbf{h}_k^H R_{\text{yy}}^{-1} \mathbf{h}_k x_k + \sum_{i \neq k} \mathbf{h}_k R_{\text{yy}}^{-1} \mathbf{h}_i x_i + \mathbf{h}_k R_{\text{yy}}^{-1} \mathbf{n} \qquad (3.48)$$

The MMSE estimate is biased: this means that $E\left[\hat{x}_{\text{MMSE},k}|x_k\right]$ is not equal to x_k. In general, a decoder expects a signal of the form $x_k + \tilde{n}_k$, where $\tilde{n}_k$ contains the interference plus noise term and is independent from x_k. If we were to make hard decisions on the MMSE estimates, the bias would have a negative impact on performance (except for constant modulus constellations, because a scaling of the decoder input signal does not change the decision regions). However, the impact of the bias can somehow be suppressed as decoders rely on soft and not hard decisions in general.

A more critical aspect is for analysis purposes when using the output of the MMSE receiver because *the associated SNR does not predict the performance*. Let us rewrite expression (3.48) in the form $x_k + \tilde{n}_{\text{MMSE},k}$, where $\tilde{n}_{\text{MMSE},k}$ is the interference plus noise term as written in the following equation :

$$\hat{x}_{\text{MMSE},k} = x_k + \underbrace{(\mathbf{h}_k^H R_{\text{yy}}^{-1} \mathbf{h}_k - 1)x_k + \sum_{i \neq k} \mathbf{h}_k R_{\text{yy}}^{-1} \mathbf{h}_i x_i + \mathbf{h}_k R_{\text{yy}}^{-1} \mathbf{n}}_{\tilde{n}_{\text{MMSE},k}:\text{ Interference plus noise}} \qquad (3.49)$$

The decoder sees $\tilde{n}_{\text{MMSE},k}$ as the noise in the communication. Therefore, the associated SNR is equal to:

$$\text{SNR}_{\text{MMSE}}(k) = \frac{P_x}{E\left[\|\tilde{n}_{\text{MMSE},k}\|^2\right]} = \frac{P_x}{E\left[\|\hat{x}_{\text{MMSE},k} - x_k\|^2\right]} \qquad (3.50)$$

$\text{SNR}_{\text{MMSE}}(k)$ cannot be used to assess the system performance because *the signal of interest* x_k *and the noise term* $\tilde{n}_k$ *are not independent*. For example, the maximal achievable rate for data stream k does not have the usual expression $\log_2(1 + \text{SNR}_{\text{MMSE}}(k))$.

Unbiased MMSE: The bias of the MMSE receiver gives the motivation behind the design of the unbiased MMSE receiver. An unbiased receiver is defined as giving an output for stream k of the form:

$$\hat{x}_k = x_k + \tilde{n}_k \qquad (3.51)$$

where $\tilde{n}_k$ is independent from x_k. The ZF receiver is an example of an unbiased receiver.

Among all the unbiased receivers, the unbiased MMSE receiver is the one with the lowest MSE though the MSE is higher than the one of the MMSE receivers but the bias is removed. It turns out that this receiver has a very simple form: $\hat{x}_{\text{UMMSE},k}$ is obtained by dividing $\hat{x}_{\text{MMSE},k}$ by the bias $\mathbf{h}_k^{\text{H}} R_{\text{yy}}^{-1} \mathbf{h}_k$:

$$\hat{x}_{\text{UMMSE},k} = x_k + \underbrace{\sum_{i \neq k} \mathbf{h}_k R_{\text{yy}}^{-1} \mathbf{h}_i x_i / \mathbf{h}_k^{\text{H}} R_{\text{yy}}^{-1} \mathbf{h}_k + \mathbf{h}_k R_{\text{yy}}^{-1} \mathbf{n} / \mathbf{h}_k^{\text{H}} R_{\text{yy}}^{-1} \mathbf{h}_k}_{\tilde{n}_{\text{UMMSE},k}: \text{ Interference plus noise}} \qquad (3.52)$$

We can verify that the interference plus noise term $\tilde{n}_{\text{UMMSE},k}$ is independent from x_k. Hence, the unbiased MMSE receiver is:

$$F_{\text{UMMSE}}(k) = \frac{\mathbf{h}_k^{\text{H}} R_{\text{yy}}^{-1}}{\mathbf{h}_k^{\text{H}} R_{\text{yy}}^{-1} \mathbf{h}_k} \qquad (3.53)$$

Furthermore, the post-processing SNR associated with stream k is defined as:

$$\text{SNR}_{\text{UMMSE}}(k) = \frac{P_x}{E\left[\|\tilde{n}_{\text{UMMSE},k}\|^2\right]} = \frac{P_x}{E\left[\|\hat{x}_{\text{UMMSE},k} - x_k\|^2\right]} \qquad (3.54)$$

Some algebraic manipulations allow for an alternative expression of the receiver and a simple expression for the post-processing SNR.

The expression of the unbiased MMSE receiver associated with stream k is:

$$F_{\text{UMMSE}}(k) = \frac{\mathbf{h}_k^{\text{H}} \bar{R}_k^{-1}}{\mathbf{h}_k^{\text{H}} \bar{R}_k^{-1} \mathbf{h}_k} \qquad \text{where} \quad \bar{R}_k = \sum_{i=1, i \neq k}^{M_{\text{T}}} P_x \mathbf{h}_i \mathbf{h}_i^{\text{H}} + \sigma_n^2 I \qquad (3.55)$$

The post-processing SNR is:

$$\text{SNR}_{\text{UMMSE}}(k) = \mathbf{h}_k^{\text{H}} \bar{R}_k^{-1} \mathbf{h}_k \qquad (3.56)$$

A block-wise expression of the unbiased MMSE receiver can be conveniently used as well, based on the expression of the MMSE receiver (3.47). The vectorial output of the MMSE receiver is:

$$\hat{\mathbf{x}}_{\text{UMMSE}} = F_{\text{MMSE}} \mathbf{y} = F_{\text{MMSE}} H \mathbf{x} + F_{\text{MMSE}} \mathbf{n} \qquad (3.57)$$

Instead of an individual bias per stream, we determine the block-wise bias which is made of the diagonal elements of the matrix $F_{\text{MMSE}} H$. Removing the bias is equivalent to inverting the diagonal elements of $F_{\text{MMSE}} H$.

A block-wise expression of the unbiased MMSE receiver is:

$$F_{\text{UMMSE}} = \mathcal{D}^{-1} \left(H^{\text{H}} H + \frac{\sigma_n^2}{P_x} I \right)^{-1} H^{\text{H}} \tag{3.58}$$

The matrix $\mathcal{D}$ is a diagonal matrix whose diagonal is equal to the diagonal of $\left(H^{\text{H}} H + \frac{\sigma_n^2}{P_x} I \right)^{-1} H^{\text{H}} H$. The post-processing SNR for stream k is:

$$\text{SNR}_{\text{UMMSE}}(k) = \frac{P_x}{\sigma_n^2 \left[K_{\text{MMSE}}^{-1} \right]_{k,k}} - 1, \text{ with } K_{\text{MMSE}} = H^{\text{H}} H + \frac{\sigma_n^2}{P_x} I \tag{3.59}$$

Based on the expression (3.59), the asymptotic behaviour of the MMSE receiver can be extracted at low and high SNR $\rho = \bar{P}/\sigma_n^2$ for a fixed value of the channel matrix: (a) at high SNR, the MMSE becomes equivalent to the ZF receiver and (b) at low SNR, the MMSE becomes equivalent to the MF receiver.

3.1.6.1 Geometric Interpretation

In the ZF framework, the interfering data is treated as an unknown deterministic quantity. In the MMSE context, the interfering data is modelled as a random Gaussian variable. Considering the estimation of symbol x_k, we write the received signal as:

$$\mathbf{y} = \mathbf{h}_k x_k + \sum_{i \neq k} \mathbf{h}_i x_i + \mathbf{n} \tag{3.60}$$

The interference plus noise term is $\sum_{i \neq k} \mathbf{h}_i x_i + \mathbf{n}$ is modelled as a coloured Gaussian noise with covariance matrix:

$$\bar{R}_k = P_x \sum_{i \neq k} \mathbf{h}_i \mathbf{h}_i^{\text{H}} + \sigma_n^2 I \tag{3.61}$$

(3.60) defines a SIMO channel with coloured noise for which we have described the optimal receiver in Section 3.17. Hence, the MMSE receiver is interpreted as a 2-step operation: (1) whitening of the interference plus noise term and (2) matched filtering as follows:

Whitening: The interference plus noise term is whitened by multiplying the received signal by $\bar{R}_k^{-1/2}$ as:

$$\bar{R}_k^{-1/2} \mathbf{y} = \bar{R}_k^{-1/2} \mathbf{h}_k x_k + \bar{R}_k^{-1/2} \left(\sum_{i \neq k} \mathbf{h}_i x_i + \tilde{\mathbf{n}} \right) \tag{3.62}$$

This new system is equivalent to a SIMO system with channel $\bar{R}_k^{-1/2} \mathbf{h}_k$ and white Gaussian noise. Then, we know that the optimal processing is spatial matched filtering.

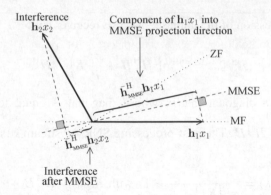

Figure 3.7 MMSE operation for a 2×2 system to estimate stream 1 with channel $\mathbf{h}_1$ the received signal is projected into a direction that is in-between the MF and the ZF projection direction.

Matched Filtering: The filter matched to $\bar{\boldsymbol{R}}_k^{-1/2}\mathbf{h}_k$ is $\mathbf{h}_k^{\mathrm{H}}\bar{\boldsymbol{R}}_k^{-H/2}$. Noting that $\bar{\boldsymbol{R}}_k^{-H/2}\bar{\boldsymbol{R}}_k^{-1/2} = \bar{\boldsymbol{R}}_k^{-1}$, the signal after MF is:

$$\hat{x}_{\mathrm{MMSE},k} = (\mathbf{h}_k^{\mathrm{H}}\bar{\boldsymbol{R}}_k^{-1}\mathbf{h}_k)\,x_k + \mathbf{h}_k^{\mathrm{H}}\bar{\boldsymbol{R}}_k^{-1}\mathbf{n} \tag{3.63}$$

Illustrative Example: We illustrate the MMSE operation for a 2×2 system with real channels in Figure 3.7. The received signal is $\mathbf{h}_1 x_1 + \mathbf{h}_2 x_2 + \mathbf{n}$. The purpose is to detect stream 1 with channel $\mathbf{h}_1$. Stream 2 with channel $\mathbf{h}_2$ is an interferer. The received signal is projected into a direction that is in-between the MF and the ZF projection direction. As a result, the output of the MMSE receiver contains a interference component which does not exist in the ZF output, and the signal gets distorted, which does not happen in the MF output.

Unbiased MMSE receiver

The MMSE receiver is the receiver minimising the MSE, that is, $E_{\mathbf{x},\mathbf{n}}\|\hat{\mathbf{x}} - \mathbf{x}\|^2$. Because the MMSE receiver is biased, we defined the unbiased MMSE receiver among all the unbiased receivers as one with the highest MSE. Denoting $\boldsymbol{K}_{\mathrm{MMSE}} = \boldsymbol{H}^{\mathrm{H}}\boldsymbol{H} + \frac{\sigma_n^2}{P_x}\boldsymbol{I}$ and $\mathcal{D} = \mathrm{diag}(\boldsymbol{K}_{\mathrm{MMSE}}^{-1}\boldsymbol{H}^{\mathrm{H}}\boldsymbol{H})$, the formulas of interest are summarised in the following table.

	Filter	Post-processing SNR
Block-wise	$\boldsymbol{F}_{\mathrm{UMMSE}} = \mathcal{D}^{-1}\boldsymbol{K}_{\mathrm{MMSE}}^{-1}\boldsymbol{H}^{\mathrm{H}}$	$\mathrm{SNR}_{\mathrm{UMMSE}}(k) = \dfrac{P_x}{\sigma_n^2[\boldsymbol{K}_{\mathrm{MMSE}}^{-1}]_{kk}} - 1$
Stream-wise	$\boldsymbol{F}_{\mathrm{UMMSE}}(k) = \dfrac{\mathbf{h}_k^{\mathrm{H}}\bar{\boldsymbol{R}}_k^{-1}}{\mathbf{h}_k^{\mathrm{H}}\bar{\boldsymbol{R}}_k^{-1}\mathbf{h}_k}$	$\mathrm{SNR}_{\mathrm{UMMSE}}(k) = P_x\mathbf{h}_k^{\mathrm{H}}\bar{\boldsymbol{R}}_k^{-1}\mathbf{h}_k$

3.1.7   SIC Receiver and V-Blast

The SIC receiver is an iterative receiver. The principle of SIC is to decode one stream of data at a time and remove its contribution from the received signal:

- At each iteration, 1 stream of data is decoded considering the other streams as interference.
- The decoded stream is then removed from the list of interfering streams.
- At the following iteration, the selected stream is decoded with one less interferer. Each iteration reduces the number of interfering stream by one.

3.1.7.1   SIC Receiver

Following the notations defined in Section 3.1.1, the received signal at a given time is:

$$\mathbf{y} = \mathbf{h}_1 x_1 + \mathbf{h}_2 x_2 + \cdots + \mathbf{h}_{M_T} x_{M_T} + \mathbf{n} \qquad (3.64)$$

The received signal is made out of contributions from all transmitted data streams that can be viewed as *layers*. The word of *layers* is frequently used for SIC receivers to designate the transmitted data streams. This terminology was introduced in the context of V-Blast whose receiver relies on successive interference cancellation. The received signal consists of the addition of M_T layers. The top layer is first decoded considering the other layers as interference. Once a layer is decoded, its contribution is removed and the next layer is decoded.

The different steps of the SIC receivers are now described. To simplify the presentation, we describe the operations on a symbol basis.

Step 1

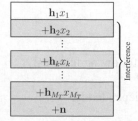

- **Layer decoding:** Using a linear receiver, layer 1 (codeword containing x_1) is decoded based on the received signal $\mathbf{y}_{L_1} = \mathbf{y}$ considering streams 2 to M_L as interferers:

$$\mathbf{y}_{L_1} = \mathbf{h}_1 x_1 + \underbrace{\sum_{i=2}^{M_T} \mathbf{h}_i x_i}_{\text{Interference}} + \mathbf{n}.$$

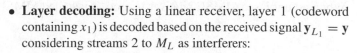

- **Layer removal:** The contribution of layer 1, that is, $\mathbf{h}_1 x_1$, in the signal $\mathbf{y}_{L_1}$ is reconstructed and removed from the signal $\mathbf{y}_{L_1}$. Layer 2 is decoded based on the new signal $\mathbf{y}_{L_2} = \mathbf{y}_{L_1} - \mathbf{h}_1 x_1$.

Step k

- **Layer decoding:** Using a linear receiver, layer k is decoded based on the signal $\mathbf{y}_{L_k}$:

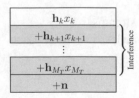

$$\mathbf{y}_{L_k} = \mathbf{y} - \underbrace{\sum_{i=1}^{k-1}\mathbf{h}_i x_i}_{\substack{\text{Interference removed}\\\text{at previous iterations}}} = \mathbf{h}_k x_k + \underbrace{\sum_{i=k+1}^{M_T}\mathbf{h}_i x_i}_{\substack{\text{Interference}\\\text{at current iteration}}} + \mathbf{n}.$$

In $\mathbf{y}_{L_k}$, streams 1 to $k-1$ have been removed during previous iterations. In the decoding of layer k, layers $k+1$ to M_T are interferers.

- **Layer removal:** The contribution of layer k, that is, $\mathbf{h}_k x_k$, in the signal $\mathbf{y}_{L_k}$ is reconstructed and removed from $\mathbf{y}_{L_k}$. The new signal from which layer $k+1$ is decoded is $\mathbf{y}_{L_{k+1}} = \mathbf{y}_{L_k} - \mathbf{h}_k x_k$.

Step M_T

- **Layer decoding:** All interfering layers have been removed. Decoding of stream M_T is based on the interference free signal:

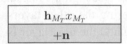

$$\mathbf{y}_{L_{M_T}} = \mathbf{h}_{M_T} x_{M_T} + \mathbf{n}.$$

This is a SIMO system for which the optimal receiver is the spatial matched filter.

At step k, a linear receiver is used to get a soft estimate of all the symbols of codeword k. Based on all the soft estimates, the codeword is decoded after which hard estimates of all the symbols are obtained. When a ZF receiver is used, the receiver is called ZF-SIC. When an unbiased MMSE receiver is used, it is called MMSE-SIC.

3.1.7.2   V-Blast

The V-Blast architecture is depicted in Figure 3.8. It relies on the following features:

1. The input streams transmitted from each antenna are coded independently.
2. The receiver is based on SIC.

When the decoding of each layer relies on the unbiased MMSE (UMMSE) receiver, V-Blast achieves the maximal rate for a reliable communication. Under fast fading conditions, the

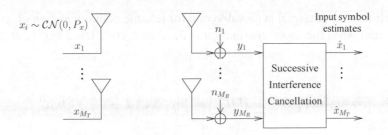

Figure 3.8   V-Blast architecture.

V-Blast architecture is optimal and achieves the capacity. Those points are detailed in the next section.

3.1.7.3   Capacity Achieving Structure for the Fast Fading MIMO Channel

Without loss of generality, we consider the first step of the SIC receiver, with a two-stage operation: (1) apply the UMMSE receiver to recover the first layer and (2) remove the reconstructed layer from the MIMO system: the new MIMO system has a reduced number of $M_T - 1$ unknown input symbols. *This two-stage operation is information lossless.* More precisely, the maximal achievable rate for reliable communication of the initial $M_T \times M_R$ MIMO system is equal to the sum of (a) the maximal achievable rate at the output of the UMMSE for stream 1 and (b) the maximal achievable rate of the reduced sized $(M_T - 1) \times M_R$ MIMO system.

Let us look more closely at the maximal achievable rate for the two stages previously mentioned:

1. From Section 3.1.6, the SNR for stream 1 at the output of the UMMSE is

$$SNR(L_1) = P_x \mathbf{h}_1^H \bar{\mathbf{R}}_1^{-1} \mathbf{h}_1. \tag{3.65}$$

where $\bar{\mathbf{R}}_1 = P_x \sum_{i>1} \mathbf{h}_i \mathbf{h}_i^H + \sigma_n^2 \mathbf{I}$. We consider a scenario where the transmission parameters for layer 1 are based on the first UMMSE output or in other words on $SNR(L_1)$. Then, the maximal achievable rate for layer 1 is:

$$\mathcal{R}_{\max}(L_1) = \log_2 \det [1 + SNR(L_1)] \tag{3.66}$$

2. Once stream 1 is removed, the system becomes equivalent to a new MIMO system with $M_T - 1$ unknown input symbols and M_R outputs. This system has a maximal achievable rate for reliable communication equal to:

$$\bar{\mathcal{R}}_{\max}(L_1) = \log_2 \det \left(\mathbf{I} + \frac{P_x}{\sigma_n^2} \sum_{i>1} \mathbf{h}_i \mathbf{h}_i^H \right) = \log_2 \det \left(\bar{\mathbf{R}}_1 / \sigma_n^2 \right). \tag{3.67}$$

We recall that the maximal achievable rate for reliable communication of an $M_T \times M_R$ MIMO system when the input distribution is fixed with covariance matrix $R_{xx} = P_x I$ is equal to:

$$\mathcal{R}_{\max} = \log_2 \det \left(I + \frac{P_x}{\sigma_n^2} H H^H \right) = \log_2 \det \left(I + \frac{P_x}{\sigma_n^2} \sum_{i=1}^{M_R} \mathbf{h}_i \mathbf{h}_i^H \right) \quad (3.68)$$

We conveniently rewrite $\mathcal{R}_{\max}$ as $\log_2 \det(R_{yy}/\sigma_n^2)$, where $R_{yy} = \sigma_n^2 I + P_x \sum_{i=1}^{M_R} \mathbf{h}_i \mathbf{h}_i^H$ is the covariance matrix of the received signal $\mathbf{y}$. Using some algebraic manipulations, we can prove the following:

$$\begin{aligned} \log_2 \det(R_{yy}/\sigma_n^2) &= \log_2 \det(1 + \text{SNR}(L_1)) + \log_2 \det(\bar{R}_1/\sigma_n^2) \\ &= \qquad \mathcal{R}_{\max}(L_1) \qquad\qquad + \qquad \bar{\mathcal{R}}_{\max}(L_1) \end{aligned} \quad (3.69)$$

This result is also valid for any step of the SIC receiver as described below.

Each step of the MMSE-SIC is information lossless: the maximal achievable rate of the system with k interfering layers is equal to the sum of the maximal achievable rate of the kth stream after the UMMSE receiver and the maximal achievable rate of the reduced size $(M_T - k) \times M_R$ MIMO system after removal of the kth stream. This equality is expressed in mathematical terms as:

$$\begin{aligned} \log_2 \det(\bar{R}_k/\sigma_n^2) &= \log_2 \det(1 + \text{SNR}(L_k)) + \log_2 \det(\bar{R}_{k+1}/\sigma_n^2) \\ &= \qquad \mathcal{R}_{\max}(L_k) \qquad\qquad + \qquad \bar{\mathcal{R}}_{\max}(L_k) \end{aligned} \quad (3.70)$$

$\text{SNR}(L_k) = P_x \mathbf{h}_k^H \bar{R}_k^{-1} \mathbf{h}_k$, $\bar{R}_k = P_x \sum_{i>k} \mathbf{h}_i \mathbf{h}_i^H + \sigma_n^2 I$. This relationship leads to the following result:

$$\mathcal{R}_{\max} = \sum_{i=1}^{M_T} \mathcal{R}_{\max}(L_i) = \sum_{i=1}^{M_L} \log_2 \left[1 + \text{SNR}(L_i) \right]. \quad (3.71)$$

The maximal achievable rate for reliable communication is the sum of the maximal achievable rates for each layer of the MMSE-SIC receiver.

Expression (3.71) leads to the optimality of V-Blast in fast fading channels, as:

$$E \left[\log_2 \det(R_{yy}/\sigma_n^2) \right] = \sum_{i=1}^{M_T} E \left[\log_2 \det \left[1 + \text{SNR}(L_i) \right] \right] \quad (3.72)$$

The capacity of a fast fading channel is $E \left[\log_2 \det(R_{yy}/\sigma_n^2) \right]$ while the capacity of each step of V-Blast is $E \left[\log_2 \det \left[1 + \text{SNR}(L_i) \right] \right]$.

3.1.7.4 Ordered SIC Receiver

When operating under asymptotic conditions such as very long codewords, the error probability can be brought very close to zero as long as the transmission rate is below capacity. As a result, the layers in SIC can be decoded in any order. However, in practice those conditions are not met: errors will happen at each stage of the SIC receiver and might propagate within layers. To minimise the risk of error propagation, the decoding of layers can be ordered. If the same rate is selected for each stream, a good strategy is to order the stream based on the post-processing SNR, beginning with the stream with higher SNR. Step k of the ordered MMSE-SIC based on SNR ordering is as follows.

Step k

- **Stream ordering:** The post-processing SNR is computed for each stream $i = k, \ldots, M_T$ as:

$$\text{SNR}^{(k)}(L_i) = \frac{P_x}{\sigma_n^2 R_{ii}^{(k)}} - 1$$

where $R_{ii}^{(k)}$ is the ith diagonal element of matrix $R^{(k)} = \left(\bar{H}^{(k)\text{H}} \bar{H}^{(k)} + \sigma_n^2/P_x I \right)^{-1}$. $R^{(k)}$ is a square matrix of dimension $M_T - k + 1$, that is, the number of remaining streams to be estimated at step k.

  $\bar{H}^{(k)} = \begin{bmatrix} \mathbf{h}_k & \mathbf{h}_{k+1} \ldots \mathbf{h}_{M_T} \end{bmatrix}$ is the equivalent channel corresponding to the transmission of streams $i = k, \ldots, M_T$.

- **Stream re-indexing**: The post-processing SNRs $\{\text{SNR}_{L_i}^{(k)}, i = k \cdots M_T\}$ are sorted in descending order. Input streams and all related quantities are re-indexed from k to M_T.

- **Stream decoding**: The stream with highest SNR value (re-indexed as k) is decoded based on the signal $\mathbf{y}_{L_k}$:

$$\mathbf{y}_{L_k} = \mathbf{h}_k x_k + \sum_{i=k+1}^{M_T} \mathbf{h}_i x_i + \mathbf{n}.$$

- **Stream removal**: The contribution of stream k, that is, $\mathbf{h}_k x_k$, in the signal $\mathbf{y}_{L_k}$ is reconstructed and removed from $\mathbf{y}_{L_k}$. The new signal from which stream $k + 1$ is estimated is $\mathbf{y}_{L_{k+1}} = \mathbf{y}_{L_k} - \mathbf{h}_k x_k$.

For performance optimisation, the rate for each input layer should be adapted to the corresponding post-processing SNR to maximise the total throughput. Hence, the layers with higher post-processing SNR use a higher rate. In a practical system, the error probability might not be negligible even when operating below capacity. Thus, the error probability might become too low when using a rate that is too high, Therefore, the selected rates should also result in a low probability of error. In summary, a proper design should have two steps: (1) ordering of the layers according to the post-processing SNR and (2) selection of an appropriate rate for each layer with a low error probability in order to minimise the risk of error propagation.

Ordered SIC: Numerical Example: Consider the following 3×3 channel matrix carrying 3 streams of data of SNR $\rho = \bar{P}/\sigma_n^2 = 10$ dB:

$$H = \begin{bmatrix} 0.67 - j0.64 & 0.22 - j0.29 & -0.46 + j0.43 \\ 0.07 - j0.20 & -0.83 - j0.36 & -0.87 + j0.04 \\ -0.79 - j0.33 & -0.68 + j0.87 & -0.19 - j1.04 \end{bmatrix}. \qquad (3.73)$$

In the table below, we describe the different steps of ordered MMSE-SIC. The SNR in the table is the SNR at the output of the UMMSE for each stream to be estimated.

	Step 1			Step 2			Step 3
Original ordering	SNR per stream	New index	Stream decoded	SNR per stream	New index	Stream decoded	Stream decoded
1	2.89	3	×	3.17	3	×	3
2	4.08	2	×	4.32	2	2	×
3	6.59	1	1	×	1	×	×

Successive interference cancellation

The SIC receiver is an iterative receiver where each stream is decoded successively and its contribution removed from the received signal:

▶ Stream 1 is decoded considering streams 2 to M_T as interferers. Once decoded, its contribution is removed from the received signal
▶ When decoding stream k, the interference from streams 1 to $k - 1$ has been removed during previous iterations. Decoding is done considering streams $k + 1$ to M_T as interferers. The corresponding post processing SNR for stream k is:

$$\text{SNR}(L_k) = P_x \mathbf{h}_k^H \bar{R}_k^{-1} \mathbf{h}_k, \qquad \text{with: } \bar{R}_k = P_x \sum_{i>k} \mathbf{h}_i^H \mathbf{h}_i^H + \sigma_n^2 I \qquad (3.74)$$

The V-Blast architecture is based on the following features:

▶ The input streams are coded independently.
▶ The receiver is based on SIC.

Successive interference cancellation (Continued)

The maximal achievable rate of the architecture in Figure 3.1 is achieved by V-Blast: a ML receiver is not necessary for optimality as the MMSE-SIC receiver brings optimal performance. The maximal achievable rate can be written as a function of $\text{SNR}(L_k)$ as:

$$\mathcal{R}_{\text{max}} = \log_2 \det \left[I + \frac{P_x}{\sigma_n^2} HH^{\text{H}} \right] = \sum_{k=1}^{M_T} \log_2 \det [1 + \text{SNR}(L_k)] \tag{3.75}$$

V-Blast achieves capacity for fast fading channels.

3.1.8   *Performance*

3.1.8.1   **Performance Measures**

We want to assess the performance of a MIMO system where both the distribution of the input symbols (i.i.d.) and the receiver is fixed. We assume an optimisation strategy of the input parameters where the rate and coding of each input stream are adjusted independently: the rate and coding of stream k is based on the corresponding SNR obtained after the receiver $\text{SNR}(k)$ (or k^{th} layer for SIC).

3.1.8.2   **Time-invariant and Fast Fading Channel: Maximal Achievable Rate**

For each stream, the maximal achievable rate for reliable communication is $\log_2 [1 + \text{SNR}(k)]$. This rate is achieved when input stream k has a Gaussian distribution and the CSI is known at the transmission. Hence, the total achievable rate for communication is the sum of maximal achievable rates over all the streams:

$$\mathcal{R}_{\text{max}} = \sum_{k=1}^{M_T} \log_2 [1 + \text{SNR}(k)] \tag{3.76}$$

When the channel is fast fading, the maximal achievable rate is $E[\mathcal{R}_{\text{max}}]$. As stated in Section 3.1.7, the maximal achievable rate evaluated for the SIC receiver is also equal to the capacity of a fast fading i.i.d. Rayleigh channel.

Slow Fading Channels: Outage Probability per Stream: The relevant outage measure is the outage probability *per stream*. Indeed, computing the overall outage probability, as defined in Chapter 2, based on the sum rate in (3.76) is not adequate: the transmission structure minimising the corresponding outage probability involves coding across antennas, while we assume that the input streams are independent from each other. Hence, such a global outage probability does not measure the performance of the actual structure of the transceivers studied in this section. The outage probability for stream k for selected rate R is computed as:

$$p_{\text{out}}(k) = \text{Pr} \left\{ \log_2 [1 + \text{SNR}(k)] < \text{R} \right\} \tag{3.77}$$

The outputs of the linear receivers have the same statistics. Therefore, we do not distinguish them. However, the outage probability of each output of a SIC receiver is different and needs to be computed individually.

3.1.8.3 Multiplexing Gain

To determine the multiplexing gain of the receivers, we examine their behaviour at high SNR $\rho = \bar{P}/\sigma_n^2$. The maximal achievable rate of the receivers (except MF) behaves asymptotically as:

$$\mathcal{R}_{\max} \sim r_H \log_2 \rho \tag{3.78}$$

$$E\left[\mathcal{R}_{\max}\right] \sim \min(M_T, M_R) \log_2 \rho. \tag{3.79}$$

Hence, for the ZF, MMSE, and SIC receivers, using definition (2.97), the multiplexing gain is equal to the rank of the channel matrix r_H when transmission is done over a fixed channel realisation. For a fading channel, the multiplexing gain is equal to $\min(M_T, M_R)$. Even if ZF and MMSE receivers are not optimal (SIC is optimal), they still exhibit the maximal multiplexing gain.

3.1.8.4 Diversity Gain

Consider the ZF receiver in the noise free case. In order to estimate a given stream of data, the inter-stream interference is cancelled by projecting the received signal into the space orthogonal to the interference space. The multi-stream signal Hx lies in a space of dimension equal to M_T, that is, the dimension of the column space of H. After the projection, each stream lies in a space of reduced dimension due to the projection operation. The dimension of this space gives the degrees of freedom still available after the projection for the estimation of each stream. Intuitively speaking, those degrees of freedom can also be interpreted as the diversity gain.

Considering the estimation of a given stream, the dimension of the interference space is equal to $M_T - 1$. Hence the dimension of the space orthogonal to the interference is equal to $M_R - M_T + 1$. After the projection, one given output stream lies in a space of dimension $M_R - M_T + 1$. This dimension limits the diversity gain which is equal to $M_R - M_T + 1$. As the MMSE receiver is equivalent to the ZF receiver at high SNR, it exhibits the same diversity gain per stream $M_R - M_T + 1$.

Consider now the SIC receiver. Although we are in the context of slow fading channels, we assume that there is no error propagation, that is, streams are perfectly removed at each stage of the SIC receiver. In reality, this is not true. However, this assumption allows for a simple analysis which gives some insight into the mechanisms behind of the SIC receiver. At each stage of SIC, one stream of data is removed. Hence, at each stage, the diversity gain is increased. It is equal to $M_R - M_T + k$ for layer k.

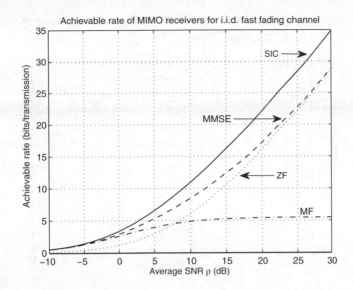

Figure 3.9 Achievable rate of MIMO receivers for a fast fading channel. SIC achieves capacity. At high SNR, ZF and MMSE receivers become equivalent. At low SNR, the ZF receiver performs the worse.

3.1.8.5 Numerical Evaluations

We present some performance aspects of the receivers according to $\rho = \bar{P}/\sigma_n^2$, for fast and slow fading channels. The channel is assumed to be i.i.d. Rayleigh fading with $E\left[|h_{ij}|^2\right] = 1$. The MIMO systems considered have 4 transmit and 4 receive antennas. The MISO/SIMO systems 4 transmit/receive antennas and the power per stream is $P_x = \bar{P}/4$.

Fast Fading Channel: Figure 3.9 shows the maximal achievable rate for reliable communication of the different receivers. From the figure, we can make the following observations:

- *At high SNR*: The ZF and MMSE receivers become equivalent. The performance of MF, limited by the inter-stream interference, is poor. The SIC receiver performs the best: it achieves capacity as expression (3.76) is the capacity for an i.i.d. Rayleigh fast fading channel.
- *At low SNR*: The performance of the different receivers appears close, except for the ZF receiver which performs the worse.

Slow Fading Channel: The ZF and MMSE receivers exhibit a dissimilar asymptotic behaviour when the channel is slowly fading. For both receivers, the outage probability of an individual stream (the outage probability is the same for each stream) is plotted in Figure 3.10 with respect to ρ. The selected rates per stream are $R = 2, 4, 6$ bits per transmission.

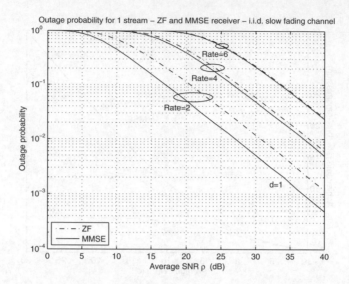

Figure 3.10 Outage probability of ZF and MMSE receivers for a slow fading channel for rate per stream equal to 2, 4 and 6 bits per transmission. ZF and MMSE receivers are not equivalent.

Next, in Figure 3.11, we compare the outage probability for each layer of the nonordered MMSE-SIC (which does not account for error propagation). Recalling that the diversity gain per layer is $M_R - M_T + k$ where k is the layer number, we observe in the figure that the diversity gain increases from 1 for the first layer to 4 for the last layer.

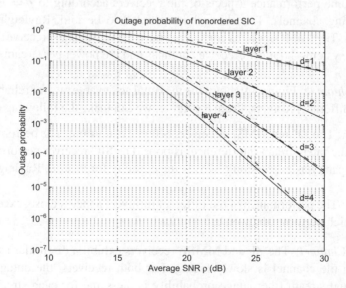

Figure 3.11 Outage probability of nonordered SIC. Slope at high SNR indicates the diversity gain.

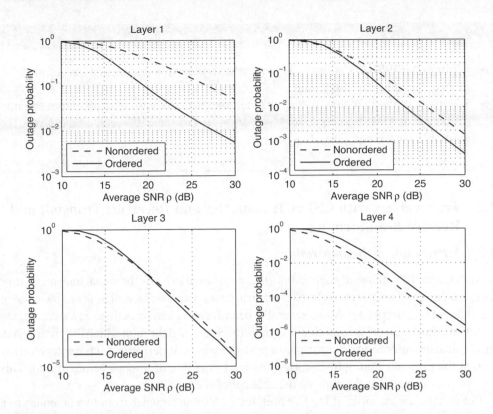

Figure 3.12 Outage probability of nonordered and ordered SIC. First layers of ordered SIC have a lower outage probability.

At last, in Figure 3.12, we compare the outage probability per layer of nonordered MMSE-SIC and ordered MMSE-SIC where the ordering is based on the post-processing SNR. Layer ordering significantly improves the outage performance of the first layers, which are the most critical layers in terms of error propagation. The outage probability of the last ordered layers might actually be higher than that of nonordered SIC, but those layers are less critical for error propagation. Note that nonordered and ordered MMSE-SIC exhibit the same diversity gain.

Receivers: performance

Consider the transceiver structure in Figure 3.1 where i.i.d. streams are sent where ZF, MMSE and SIC receivers are used.

▶ The multiplexing gain is equal to r_H for a fixed channel realisation. For a fading channel, the multiplexing gain is equal to $\min(M_T, M_R)$. Hence, the maximal multiplexing gain can be achieved even if the optimal receiver is not used.

Receivers: performance (Continued)

▶ While MMSE-SIC is optimal, ZF and MMSE receivers results in a loss of performance due to loss in array gain.

▶ When operating in a slow fading channel, the diversity gain of each output stream of the ZF or MMSE is equal to $M_T - M_R + 1$. For the SIC receiver, the diversity gain of layer k is equal to $M_T - M_R + k$.

3.2 Transceivers with CSI at Transmitter and Receiver: Transmit and Receive Beamforming

3.2.1 Principle of Beamforming

Beamforming is traditionally defined as a technique used to focus the signal transmission or reception in a certain direction. It relies on an array of antennas as well as the knowledge of the channel (e.g. its phase). More generally, beamforming can be defined as a technique to match the transmission or reception of signals to the channel through which they propagate. In a typical beamforming design, a pre-or post-processing is performed at the antenna array so that the post-processing SNR corresponding to the link of interest is maximised. This pre- or post-processing depends on the channel of the link.

Let us take the example of the transmission of a same signal $\tilde{x}$ from two antennas to a receiver with a single antenna as shown in Figure 3.13. The channel from the first antenna is a complex number that we write as $h_1 = |h_1|e^{j\phi_1}$. The channel from the second antenna is $h_2 = |h_2|e^{j\phi_2}$. The signals from both transmit antennas get added up at the receiver as $\left(|h_1|e^{j\phi_1} + |h_2|e^{j\phi_2}\right)\tilde{x}$. Depending on the values of the phases ϕ_1 and ϕ_2, the signals might add up destructively or constructively, as depicted in Figure 3.14:

- If $\phi_2 = -\phi_1$, the addition of the signals results in $(|h_1| - |h_2|)\tilde{x}$. The addition is destructive.
- If $\phi_2 = \phi_1$, the addition of the signals results in $(|h_1| + |h_2|)\tilde{x}$. The addition is constructive.

Figure 3.13 Left figure: signal $\tilde{x}$ is sent from two antennas; the corresponding signals might add up destructively at the receiver. Right figure: the signals sent from each antenna are preconditioned to match the channel to add constructively at the receiver.

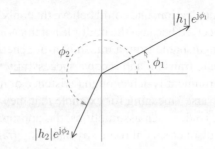

Figure 3.14 Addition of signals sent from two transmit antennas at the receive antenna of a 2×1 MISO system. Depending on the phase of the signals, the addition is constructive or destructive.

If the transmitter knows the phase of the channel, the signals can be preconditioned before being transmitted so that they add up constructively at the receiver. An additional pre-processing can be performed if the transmitter knows the amplitude of the channel as well: if the channels coefficients are known at the transmitter, the signal can be pre-processed so that the received SNR is maximised as follows.

1. *Phase alignment*: Using the phase information about the channels, the transmitted signals are pre-processed so that they add constructively at the receiver. The first antenna transmits $e^{-j\phi_1}\tilde{x}$ while the second antenna transmits $e^{-j\phi_2}\tilde{x}$. The received signals from antenna 1 and antenna 2 have the same phase (equal to zero). The corresponding received signals are $|h_1|\tilde{x}$ and $|h_2|\tilde{x}$ and add constructively as $(|h_1| + |h_2|)\tilde{x}$.
2. *Power distribution*: When the amplitude of the channel is additionally known, the transmit power can be distributed across both antennas to maximise the SNR at the receiver. This is done by allocating more power to the antenna with the stronger channel.

The linear receivers described in Section 3.1 belong to the class of receive beamforming. The kind of beamforming performed is more complex than the one presented in the example above because of the presence of multiple streams: multiple beamforming has to be performed to match each transmitted stream.

3.2.2   Multiple Transmit and Receive Beams

An array of M antennas can form up to M different beams matched to M different communication links. Each beam can carry one stream of data. A traditional configuration enabling such a multi-beam communication consists of an antenna array with antennas that are sufficiently apart communicating with devices that are spatially distinguishable. Such an antenna array can point beams towards different directions. Different beams can be formed that point to the direction of multiple users. Data can be sent to or received from those users provided that the users are sufficiently spatially separated. This configuration assumes a line of sight communication in general.

For a MIMO link communication, one could believe that only a single beam can be successfully transmitted or received because the multiple antennas at a receiver are co-located and are not spatially distinguishable from a transmitter in general. However, as described in Chapter 1, it is possible to transmit and receive successfully multiple beams, provided that (a) the scattering environment is rich enough (existence of multiple paths), (b) the antennas are sufficiently "diverse" meaning for example that they are sufficiently separated or have a distinct pattern. Under such assumptions, the appropriate transmit and receive beamforming can extract distinct spatial routes by matching transmit and receive beams.

In this section, we will go step by step, as summarised in the following table, starting by the description of different forms of single beam communications to end with multiple beam communications where multiple beams are formed at the transmitter and receiver.

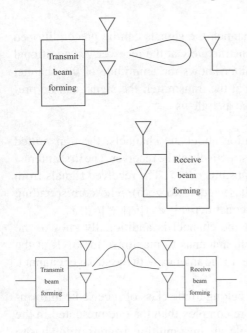

Transmit beamforming (MISO):
A transmit beamforming technique is used matching the channel from the multiple transmit antennas to the receiver.

Receive beamforming (SIMO):
A receive beamforming technique is used matching the channel from the transmit antenna to the multiple receive antennas.

Max eigenmode beamforming (MIMO):
Both transmitter and receiver form a beam matched to the underlying structure of the channel (given by the SVD of the channel matrix). Transmit and receive beams are jointly designed and match each other. They form a channel equal to the maximum singular value of the channel.

Eigenmode transmission (MIMO):
For a MIMO system, both transmitter and receiver form multiple beams matched to the underlying structure of the channel (giving by the SVD of the channel matrix). Each transmit beam is matched to a receive beam. Multiple equivalent independent channels are created equal to the singular eigenvalues of the channel. The power has to be split optimally among the channels. Eigenmode transmission requires a rich scattering environment.

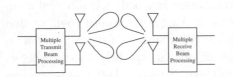

3.2.3   Transmit Beamforming (MISO System)

We consider a system with M_T transmit antennas and a single receive antenna: see Figure 3.15. We denote as $\tilde{x}$ the symbol to be sent to the receiver. Each antenna transmits a processed version of $\tilde{x}$ where its phase and amplitude are modified to match the channel to the receiver. $w_i^* \tilde{x}$ is sent from antenna i: w_i^* is a complex valued scalar weight (we use the conjugate operation for notation convenience as seen later). The weights are selected to maximise the post-processing SNR at the receiver described as follows.

3.2.3.1   Post-processing SNR

The signal $w_i^* \tilde{x}$ is sent from antenna i through channel h_i. At the receiver, the transmitted signals from all transmit antennas are added up as:

$$y = \left(\sum_{i=1}^{M_T} h_i w_i^* \right) \tilde{x} + n \tag{3.80}$$

At the receiver, the signal is seen as been transmitted from the channel $\sum_{i=1}^{M_T} h_i w_i^*$. For convenience, we rewrite this equivalent channel as the scalar product: $\mathbf{w}^H \mathbf{h}$ where $\mathbf{w} = \begin{bmatrix} w_1 & w_2 \dots w_{M_T} \end{bmatrix}^T$ is a $M_T \times 1$ row vector and $\mathbf{h} = \begin{bmatrix} h_1 & h_2 & \dots & h_{M_T} \end{bmatrix}^T$ is a column vector. The expression for the received signal becomes:

$$y = \mathbf{w}^H \mathbf{h} \, \tilde{x} + n \tag{3.81}$$

The transmit power is $\|\mathbf{w}\|^2 E|\tilde{x}|^2$ and is set to the maximal allowed transmit power denoted as $\bar{P}$: without loss of generality, we set $\|\mathbf{w}\|^2 = 1$ and $E|\tilde{x}|^2 = \bar{P}$. The power in the

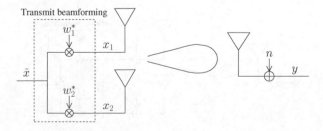

Figure 3.15   MISO System model with transmit beamforming.

signal part of (3.81) is $\bar{P}|\mathbf{w}^H\mathbf{h}|^2$, while the noise power is σ_n^2. Hence, the post-processing SNR is:

$$\mathrm{SNR} = \frac{\bar{P}|\mathbf{w}^H\mathbf{h}|^2}{\sigma_n^2} \qquad (3.82)$$

3.2.3.2   Transmit Matched Filtering

Maximising the post-processing SNR is equivalent to maximising $|\mathbf{w}^H\mathbf{h}|$, the amplitude of the inner product. As seen in Section 3.1.4, the scalar product is maximised when $\mathbf{w}$ and $\mathbf{h}$ are aligned.

The transmit filter matched to the MISO channel $\mathbf{h}$ is $\mathbf{w}_{\mathrm{TMF}} = \mathbf{h}^*/\|\mathbf{h}\|$. The Transmit MF (TMF) maximises the post-processing SNR. The received signal is $\mathbf{y}_{\mathrm{TMF}} = \|\mathbf{h}\|\tilde{x} + n$ and the post-processing SNR is:

$$\mathrm{SNR}_{\mathrm{TMF}} = \frac{\bar{P}\|\mathbf{h}\|^2}{\sigma_n^2} \qquad (3.83)$$

Transmit spatial matched filtering is also commonly called transmit Maximal Ratio Combining (MRC).

Transmit MF has a very simple structure as it involves only linear operations. It transforms a MISO system into a SISO system. This transformation does not result in any loss of information. Transmit MF is a structure achieving the capacity of a MISO system when the CSIT and CSIR are known.

3.2.4   Receive Beamforming (SIMO)

The optimal receive beamforming (Figure 3.16) is basically the symmetrical image of transmit beamforming for a MISO system. It is the spatial matched filtering already treated in Section 3.1.4.

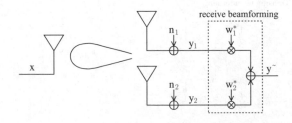

Figure 3.16   SIMO sytem with receive beamforming.

Figure 3.17 Single beam MIMO system with transmit and receive beamforming.

3.2.5 Single Beam MIMO: Maximal Eigenmode Beamforming

Multiple antennas are present at both the transmitter and receiver and both ends know the MIMO channel matrix. Then, transmit and receive beamforming can be performed jointly. We examine how to design a single transmit and receive beam carrying a single stream of data (see Figure 3.17 for 2×2 MIMO). The design criterion is the maximisation of the post-processing SNR. We denote the $M_R \times M_T$ MIMO channel matrix as $\boldsymbol{H}$, the $M_T \times 1$ transmit weight vector as $\mathbf{w}_{tx}$ and the $M_R \times 1$ receive weight vector as $\mathbf{w}_{rx}$. $\tilde{x}$ is the transmitted symbol, $\mathbf{x} = \begin{bmatrix} x_1 & x_2 & \ldots & x_{M_T} \end{bmatrix}^T$ is the vectorial input containing the signals sent from each antenna, $\mathbf{y} = \begin{bmatrix} y_1 & y_2 & \ldots & y_{M_R} \end{bmatrix}^T$ is the vectorial output containing the signals received at each antenna. The transmit power is set to $\bar{P}$. The quantities of interest are as follows.

$M_T \times 1$ transmitted signal: $\mathbf{x} = \mathbf{w}_{tx}\tilde{x}$.

where $\|\mathbf{w}_{tx}\|^2 = 1$ and $E|\tilde{x}|^2 = \bar{P}$

$M_R \times 1$ receive signal: $\mathbf{y} = \boldsymbol{H}\mathbf{w}_{tx}\tilde{x} + \mathbf{n}$

After receive beamforming: $\tilde{y} = \mathbf{w}_{rx}^T \boldsymbol{H}\mathbf{w}_{tx}\tilde{x} + \mathbf{w}_{rx}^T \mathbf{n}$

where $\|\mathbf{w}_{rx}\|^2 = 1$

After receive beamforming, the power in the signal part is $\bar{P}|\mathbf{w}_{rx}^T \boldsymbol{H}\mathbf{w}_{tx}|^2$ and the power in the noise part is $\sigma_n^2 \|\mathbf{w}_{rx}\|^2 = \sigma_n^2$. Hence, the SNR after receive beamforming is:

$$\text{SNR} = \frac{\bar{P}|\mathbf{w}_{rx}^T \boldsymbol{H}\mathbf{w}_{tx}|^2}{\sigma_n^2} \qquad (3.84)$$

Maximising the post-processing SNR is equivalent to maximising the receive power $\bar{P}|\mathbf{w}_{rx}^T \boldsymbol{H}\mathbf{w}_{tx}|^2$. The solution to this maximisation is based on the singular value decomposition of the channel matrix (see Section 2.5.3):

$$\boldsymbol{H} = \boldsymbol{U}\boldsymbol{\Lambda}\boldsymbol{V}^H = \sum_{k=1}^{r_H} \lambda_k \mathbf{u}_k \mathbf{v}_k^H \qquad (3.85)$$

$\mathbf{u}_k$ and V_k are the kth columns of U and V respectively. λ_k is the kth singular value. In addition, we conveniently denote as $\lambda_{\max}$ the maximal singular value of H, and $\mathbf{u}_{\max}$ and $\mathbf{v}_{\max}$ the corresponding left and right singular vectors. We can rewrite the term $|\mathbf{w}_{rx}^T H \mathbf{w}_{tx}|^2$ as $|\mathbf{w}_{rx}^T U \Lambda V^H \mathbf{w}_{tx}|^2$ and further decompose it as:

$$\left| \mathbf{w}_{rx}^T H \mathbf{w}_{tx} \right|^2 = \left| (\mathbf{w}_{rx}^T U \Lambda^{1/2})(\Lambda^{H/2} V^H \mathbf{w}_{tx}) \right|^2 \tag{3.86}$$

The beamforming vectors $\mathbf{w}_{tx}$ and $\mathbf{w}_{rx}$ should extract a spatial path with the highest possible energy. This is done by matching $\mathbf{w}_{tx}$ to the right singular vector of H with highest singular value and by matching $\mathbf{w}_{rx}$ to the corresponding left singular vector. The transmit beamforming vector $\mathbf{w}_{tx}$ is $\mathbf{v}_{\max}$. Hence, the product $\Lambda^{H/2} V^H \mathbf{w}_{tx}$ in Equation (3.86) is equal to $\lambda_{\max}^{1/2}$. The receive beamforming vector $\mathbf{w}_{rx}$ is $\mathbf{u}_{\max}^*$ and $\mathbf{w}_{rx}^T U \Lambda^{1/2} = \lambda_{\max}^{1/2}$.

In maximal eigenbeam transmission, both transmitter and receiver form a single beam. The transmit beamforming vector is $\mathbf{v}_{\max}$ and receive beamforming vector is $\mathbf{u}_{\max}^*$. the signal after receive beamforming is:

$$\tilde{y} = \lambda_{\max} \tilde{x} + \tilde{n} \tag{3.87}$$

$\tilde{n} = \mathbf{v}_{\max}^H \mathbf{n} \sim \mathcal{CN}(0, \sigma_n^2)$. With this transmit and receive beamforming, the system is equivalent to a SISO system with SNR equal to $\dfrac{\bar{P}\lambda_{\max}^2}{\sigma_n^2}$.

We have imposed a single beam communication. However, when the channel possesses more than one nonzero singular value, it is possible to form multiple beams and send independent data using those beams. Instead of communicating only over the channel with the maximum eigenvalue, we communicate over all the eigenchannels: such a transmission is called eigenmode transmission. We can form as many beams as non zero singular values. This approach focuses on maximising the total system throughput, while maximal eigenmode transmission focuses on a single stream communication and on maximising the post-processing SNR and hence the throughput of this single stream.

3.2.6   Eigenmode Transmission

Eigenmode transmission has already been described in Chapter 2 as a structure achieving capacity for a time-invariant channel. We give here another interpretation of this structure in beamforming terms mainly for a 2×2 MIMO system, (see Figure 3.18).

3.2.6.1   Example of the 2×2 MIMO

Let us first examine the case of two transmit and two receive antennas. Let us recall the notations linked to the singular value decomposition of H:

$$H = U \Lambda V^H \tag{3.88}$$

Figure 3.18 Eigenmode transmission: multiple beams are formed at the transmitter and receiver that match the eigenmodes of the channels.

- $\Lambda = \text{diag}(\lambda_1, \lambda_2)$ is a diagonal matrix with diagonal elements equal to λ_1 and λ_2, and $\lambda_1 \geq \lambda_2$.
- $U = [\mathbf{u}_1 \ \mathbf{u}_2]$ and $V = [\mathbf{v}_1 \ \mathbf{v}_2]$ are unitary matrices. $\mathbf{u}_1$ and $\mathbf{u}_2$ are the first and second column of U and the left singular vectors of H. $\mathbf{v}_1$ and $\mathbf{v}_2$ are the first and second column of V and the right singular vectors of H.
- $\mathbf{u}_i = \begin{bmatrix} u_{1i} \\ u_{2i} \end{bmatrix}$ and $\mathbf{v}_i = \begin{bmatrix} v_{1i} \\ v_{2i} \end{bmatrix}$.
- The SVD of H can alternatively be written as:

$$H = \lambda_1 \mathbf{u}_1 \mathbf{v}_1^{\text{H}} + \lambda_2 \mathbf{u}_2 \mathbf{v}_2^{\text{H}} \tag{3.89}$$

Transmitter: If we want to reach optimal performance, the power should be distributed optimally among the beams. This is a significant difference with the single beamforming MIMO case of Section (3.2.5). Power allocation becomes part of the optimisation problem. We denote as P_i the power assigned to beam i. The maximal overall transmit power is $\bar{P}$. Hence, the transmit power constraint is $P_1 + P_2 \leq \bar{P}$.

The first transmit beamforming vector is $\mathbf{v}_1$ and is used to send symbol $\tilde{x}_1$ ($E|\tilde{x}_1|^2 = P_1$). The second transmit beamforming vector is $\mathbf{v}_2$ and is used to send symbol $\tilde{x}_2$ ($E|\tilde{x}_2|^2 = P_2$). The transmit vector corresponding to the first beam is $\mathbf{v}_1 \tilde{x}_1$ while the transmitted vector corresponding to the second beam is $\mathbf{v}_2 \tilde{x}_2$. The beams are added up and the $M_T \times 1$ transmitted signal is:

$$\mathbf{x} = \mathbf{v}_1 \tilde{x}_1 + \mathbf{v}_2 \tilde{x}_2 \tag{3.90}$$

Receiver: Using (3.89), the received signal is:

$$\mathbf{y} = H (\mathbf{v}_1 \tilde{x}_1 + \mathbf{v}_2 \tilde{x}_2) + \mathbf{n} \tag{3.91}$$

$$= \lambda_1 \mathbf{u}_1 \tilde{x}_1 + \lambda_2 \mathbf{u}_2 \tilde{x}_2 + \mathbf{n}. \tag{3.92}$$

At the receiver, the task is to separate both streams to recover $\tilde{x}_1$ and $\tilde{x}_2$. This task is made easy because of the orthogonality properties of the received streams. From

Equation (3.92), the 2×2 equivalent channel corresponding to the transmission of $\tilde{x}_1$ and $\tilde{x}_2$ is:

$$[\lambda_1 \mathbf{u}_1 \quad \lambda_2 \mathbf{u}_2] \qquad (3.93)$$

This matrix is orthogonal because its columns are orthogonal. The optimal receiver is the spatial matched filter as explained in Section 3.1.4. To recover stream 1, the optimal receiver is the spatial filter matched to $\mathbf{u}_1$, that is, $\mathbf{u}_1{}^H$, which also eliminates the contribution of the second beam. The output of the receiver is $\tilde{y}_1 = \lambda_1 \tilde{x}_1 + \mathbf{u}_1{}^H \mathbf{n}$. Likewise, to recover stream 2, the optimal receiver is the spatial filter matched to $\mathbf{u}_2$, that is, $\mathbf{u}_2{}^H$. The output is $\tilde{y}_2 = \lambda_2 \tilde{x}_2 + \mathbf{u}_2{}^H \mathbf{n}$.

Equivalent System: The system including the transmit and receive beamforming filters becomes:

$$\tilde{y}_1 = \lambda_1 \tilde{x}_1 + \tilde{n}_1 \qquad (3.94)$$
$$\tilde{y}_2 = \lambda_2 \tilde{x}_2 + \tilde{n}_2 \qquad (3.95)$$

where $\tilde{n}_1 = \mathbf{u}_1{}^H \mathbf{n}$ and $\tilde{n}_2 = \mathbf{u}_2{}^H \mathbf{n}$. $\tilde{n}_1$ and $\tilde{n}_2$ are independent and $\tilde{n}_j \sim \mathcal{CN}(0, \sigma_n^2 \mathbf{I})$. Hence, we have created two independent spatial channels or parallel channels with respective SNRs:

$$\text{SNR}(1) = \frac{P_1 \lambda_1^2}{\sigma_n^2} \quad \text{and} \quad \text{SNR}(2) = \frac{P_2 \lambda_2^2}{\sigma_n^2} \qquad (3.96)$$

Now we still need to find the optimal power allocation.

Power Allocation through Achievable Rate Maximisation: Maximal eigenbeam transmission involves a single beam and hence a single SNR. Selecting the beamforming vectors that maximise the SNR makes perfect sense as it is equivalent to maximising the achievable rate for a reliable communication for a time invariant channel. The situation is now different because we have several parallel channels and hence more than one SNR. Both SNRs cannot be maximised at the same time: when SNR(1) increases, SNR(2) decreases. An optimisation based directly on the SNRs is not straightforward to design in general. A more relevant optimisation design relies on the achievable rate, where the optimal power allocation is the one maximising the system capacity.

Because the channels in (3.95) are independent, the achievable rate of the whole system is the sum of the capacity of the parallel channels:

$$\mathcal{R}_{\max} = \log_2 [1 + \text{SNR}(1))] + \log_2 [1 + \text{SNR}(2)] \qquad (3.97)$$

As $\bar{P} = P_1 + P_2$, we replace P_2 by $\bar{P} - P_1$ to obtain:

$$\mathcal{R}_{\max} = \log_2 \left[1 + P_1 \frac{\lambda_1^2}{\sigma_n^2} \right] + \log_2 \left[1 + (\bar{P} - P_1) \frac{\lambda_2^2}{\sigma_n^2} \right] \qquad (3.98)$$

This expression can easily be optimised with respect to P_1. The solution is as follows:

$$P_i^o = \left(\frac{1}{\gamma_0} - \frac{1}{\gamma_i} \right)^+ \tag{3.99}$$

where $x^+ = x$ if $x \geq 0$ and $x^+ = 0$ if $x < 0$. γ_0 is a value derived from the constraint $P_1^o + P_2^o = \bar{P}$. Computation of the transmit powers is done as follows:

1. First, we compute the constant γ_0 assuming that both channels are allocated power, that is, P_1^o and P_2^o are positive. We find $1/\gamma_0 = (\bar{P} + 1/\gamma_1 + 1/\gamma_2)/2$.
2. Next, we need to check that P_1^o and P_2^o computed thanks to this value of $1/\gamma_0$ are positive. We assume that $\lambda_1 \geq \lambda_2$, hence $1/\gamma_2 \geq 1/\gamma_1$ and $P_2^o \leq P_1^o$. So, we first need to check if P_2^o is positive (as it is most likely to be negative).
   - If $P_2^o = 1/\gamma_0 - 1/\gamma_2 \geq 0$, then both eigenchannels are allocated. The power allocation is as in Equation (3.99).
   - If $P_2^o = 1/\gamma_0 - 1/\gamma_2 < 0$, then $P_2^o = 0$ and $P_1^o = \bar{P}$. All the power is allocated to the first beam. In such a scenario, it is better to allocate all the power to the beam with stronger SNR.

This power allocation is called waterfilling. Waterfilling is described in further details in Section 2.5.4. Eigenmode transmission with waterfilling is a scheme achieving capacity for time invariant channels.

3.2.6.2 General MIMO Channel

The previous derivations can be generalised to a $M_T \times M_R$ MIMO system as follows.

Eigenmode transmission with waterfilling is a multiple transmit and receive beam technique where the beams are matched to the underlying structure of the channel given by the SVD of the channel matrix $H = U \Lambda V^H$:

- r_H transmit beams: $\mathbf{v}_1, \ldots, \mathbf{v}_{r_H}$.
- r_H receive beams: $\mathbf{u}_1^*, \ldots, \mathbf{u}_{r_H}^*$.
- The transmit power is allocated across eigenmodes as:

$$P_i^o = \left(\frac{1}{\gamma_0} - \frac{1}{\gamma_i} \right)^+ \tag{3.100}$$

γ_0 is the cut-off value and is determined using the power constraint :

$$\sum_{i=1}^{r_H} P_i^o = \sum_{i=1}^{r_H} \left(\frac{1}{\gamma_0} - \frac{1}{\gamma_i} \right)^+ = \bar{P} \tag{3.101}$$

Eigenmode transmission with waterfilling is an optimal structure for MIMO systems when the channel is time-invariant and is known (along with the noise variance) at the transmitter.

3.2.7 Performance of Beamforming Schemes

Performance is described according to the SNR $\rho = \bar{P}/\sigma_n^2$. The MIMO systems considered have 4 transmit and 4 receive antennas. The MISO/SIMO systems possess 4 transmit/receive antennas. The channel realisations are drawn from an i.i.d. Rayleigh fading distribution verifying $E\left[|h_{ij}|^2\right] = 1$. For MISO or SIMO systems, we plot the performance when transmit/receive spatial matched filtering is used.

3.2.7.1 Achievable Rates

In Figure 3.19, we show the achievable rates of the different beamforming schemes for a time invariant channel. The achievable rates are averaged over channel realisations drawn from an i.i.d. Rayleigh distribution. In particular, we compare eigenmode transmission with two transmit power allocations: waterfilling and uniform power allocation (equal power per beam).

• At high SNR, the curve slopes indicate the multiplexing gain for each scheme, showing a clear distinction between the schemes using MIMO capabilities and the schemes using a single stream of communication. Furthermore, as waterfilling becomes equivalent to a uniform power allocation at high SNR, eigenmode transmission with waterfilling and with uniform allocation become equivalent. Maximal eigenmode transmission outperforms MISO or SIMO transmission thanks to a larger array gain.

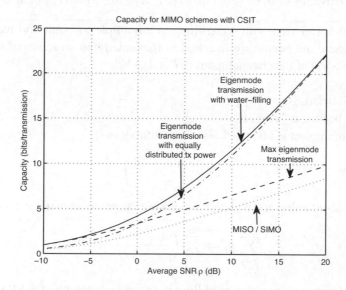

Figure 3.19 Achievable rate of beamforming schemes averaged over channel realisations drawn from an i.i.d. Rayleigh distribution.

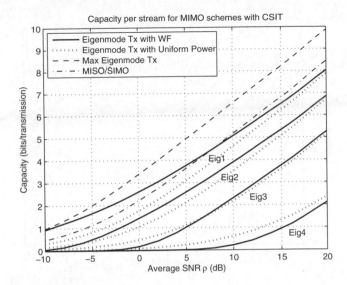

Figure 3.20 Achievable rate per stream of beamforming schemes averaged over realisations drawn from an i.i.d. Rayleigh distribution.

- At low SNR, the multiplexing capabilities are lost so eigenmode transmission becomes equivalent to maximal eigenmode transmission, while eigenmode transmission with uniform power distribution becomes equivalent to SIMO.

In Figure 3.20, the achievable rate per stream is shown. The following observations can be made: (1) at high SNR, waterfilling tends to a uniform power allocation, (2) at low to mid SNR, the stronger streams are assigned with more power and (3) at low SNR, only the strongest stream is assigned power when waterfilling is applied.

3.2.7.2   Eigenmode Transmission: Simple Example of a Practical Implementation

In the information theory framework, the codewords are assumed to be asymptotically long and the error probability asymptotically small. In practice, this assumption cannot be fulfilled and the error probability will not be negligible. Figure 3.21 shows a simple example where an uncoded QPSK constellation is used for transmission in each stream of eigenmode transmission with waterfilling. The corresponding bit error rate (BER) is shown. The slope of the uncoded BER curves indicates the diversity gain achieved by each stream.

Furthermore, capacity can take any positive value. For example, in Figure 3.19, at 5 dB the capacity of eigenmode transmission with WF is around 7.3 at 5 dB, 11.5 at 10 dB or 16.2 at 13 dB. In practice, communication systems select among a finite number of encoders and transmission rates. Hence, the curve in Figure 3.19 cannot be achieved. However, to approach capacity results, adaptive modulation and coding (AMC). is implemented. In AMC,

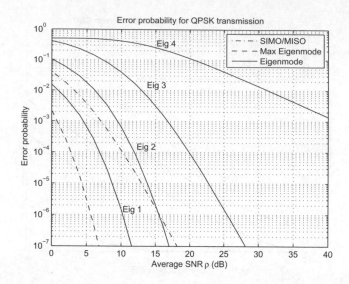

Figure 3.21   BER per stream for an uncoded QPSK constellation.

a finite number of transmission rates are available. In the present example, the system can choose between uncoded QPSK, 16-QAM, 64-QAM, hence rates of 2, 4 and 6 bits per transmission. The more AMC levels are available, the closer we can get to the optimal capacity curve. In this example, for the sake of simplicity, we assume an uncoded system, but it is of course highly desirable to have an efficient coding to achieve a low error probability.

In our example, the AMC level is selected to maximise the transmission rate under a BER constraint. The BER constraint specifies that the BER should be below 10^{-3}. In Figure 3.22, the BER curves with respect to the SNR is drawn for QPSK, 16-QAM and 64-QAM, as well as the SNR region where each constellation should be selected. In the region below 9.8 dB, the BER constraint cannot be satisfied, so the system does not transmit. In the region between 9.8 dB and 16.5 dB, the higher constellation can be transmitted verifying the BER constraint is QPSK, and similarly for the other decision regions.

After waterfilling, the post-processing SNR is computed for each stream and the appropriate AMC level is selected accordingly. The same is done for maximal eigenmode transmission and SIMO/MISO. The throughput is defined as the average number of symbols correctly decoded per transmission and is show in Figure 3.23. At high SNR, the multiplexing effect is visible: the maximal transmission rate per stream is six bits per transmission, hence, for four parallel streams, the maximal rate is 24 bits per transmission. For the single stream systems, the transmission rate cannot exceed six bits per transmission. At lower SNR, we see that maximal eigenmode transmission outperforms eigenmode transmission. Indeed, waterfilling is adapted to the maximisation of capacity and not to a transmission where only three transmission rates can be selected. To remedy this problem, two simple solutions can be implemented: (a) the number of AMC levels can be increased so

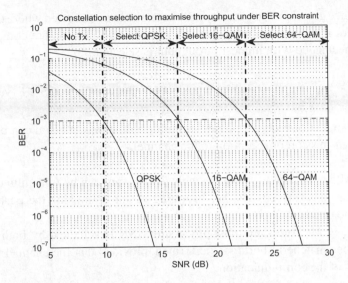

Figure 3.22   BER curves for QPSK, 16-QAM and 64-QAM and the SNR region where each constellation should be selected.

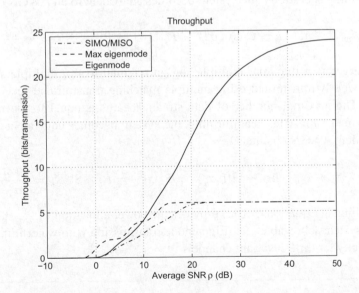

Figure 3.23   Throughput when applying a practical AMC scheme for multi-stream eigenmode transmission with WF, single stream max eigenmode transmission and SIMO/MISO.

that performance gets closer to capacity or (b) a test can be conducted where the theoretical throughputs for maximal eigenmode and eigenmode transmission are compared and the transmission technique with highest throughput is selected.

Transmit and receive beamforming

When a transmitter or receiver possesses multiple antennas, a beam can be formed to match the transmission or reception to the channel.

▶ For a SIMO or a MISO system, a single data stream can be transmitted and hence a single beam is formed. The beamforming vector optimising the post-processing SNR is the matched filter.
When $M_T = M_R$, the same post-processing SNR is achieved by both SIMO and MISO when matched filtering is performed, provided that the channel is known at both ends of the communication.
The post-processing SNR is $\bar{P}\|\mathbf{h}\|^2/\sigma_n^2$. Beamforming exhibits an array gain equal to M_T (MISO) and M_R (SIMO).

▶ For a MIMO system, when targeting a maximal array gain, a single data stream is transmitted using a single matching transmit and receive beam. The transmit beamforming vector is $\mathbf{v}_{\max}$ and the receive beamforming vector is $\mathbf{u}_{\max}^*$. With those beamforming operations, the system becomes equivalent to an AWGN channel:

$$\tilde{y} = \lambda_{\max}\tilde{x} + \tilde{n}, \quad \tilde{n} \sim \mathcal{CN}(0, \sigma_n^2), \quad E\left[|\tilde{x}|^2\right] = \bar{P}, \quad \text{SNR} = \frac{\bar{P}\lambda_{\max}^2}{\sigma_n^2} \quad (3.102)$$

▶ When targeting a maximal multiplexing gain (equal to r_H), multiple streams are simultaneously transmitted using multiple matching transmit and receive pairs of beams. The maximal number of data streams/beams is equal to the rank of the channel matrix r_H. After beamforming, the system becomes equivalent to multiple independent AWGN channels:

$$\tilde{y}_k = \lambda_k\tilde{x}_k + \tilde{n}_k, \quad \tilde{n}_k \sim \mathcal{N}(0, \sigma_n^2), \quad E\left[|\tilde{x}|^2\right] = \bar{P}, \quad \text{SNR}_k = \frac{P_k\lambda_k^2}{\sigma_n^2} \quad (3.103)$$

Water filling in Equation (3.100) provides the power allocation $P_1 \ldots P_{r_H}$ maximising the system achievable rate. Eigenmode transmission with waterfilling achieves the capacity for time-invariant channels.

3.3 Space–Time Block Codes

Space–time codes are codes that spread over time and over space (transmit antennas). They are used to improve the reliability of a wireless communication link by using transmit

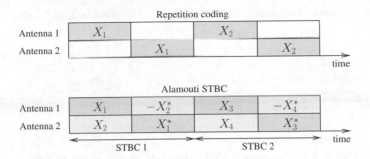

Figure 3.24 Example of an STBC based on repetition coding compared to the Alamouti STBC. The Alamouti STBC achieves twice the data rate with the same diversity gain.

diversity, that is, redundant information is sent across time and space. *Space–time codes target slow fading channels where the instantaneous CSI is not known at the transmitter.* For such a channel, we have seen in Chapter 2 that a reliable communication cannot be guaranteed. Space–time codes are useful because they reduce the outage probability compared to single antenna systems.

This section is dedicated to a special category of space–time codes called *space–time block codes*. In space–time block codes, the data is divided into blocks consisting of independent codewords. The same blocks are sent in time and space, insuring transmit diversity. Space–time block codes are advantageous because they are very simple codes requiring a low complexity encoding and decoding procedure. STBC are designed for MISO systems. They do not require multiple antennas at the receiver. In case multiple receive antennas are present, STBCs benefit from additional diversity gain as well as array gain. However, they are suboptimal for MIMO systems. Other schemes (e.g. D-Blast) achieve a higher throughput while exhibiting the same diversity gain. STBC redundancy across antennas brings diversity gain only and not coding gain. Coding gain comes from coding across each codeword (coding in time).

A very simple STBC is based on repetition coding where the same codeword is successively transmitted from a different antenna while the other antennas remain idle. It is illustrated in Figure 3.24. Repetition coding achieves the maximal diversity gain because each codeword is transmitted through all available channels. However, it is very inefficient when it comes to throughput because some antennas remain idle at each transmission time. We want STBCs that have a maximal diversity gain while still achieving a high throughput. In other words, a desired property is the minimisation of the system outage probability. Such a property cannot always be fulfilled except for the Alamouti STBC designed for a two transmit antenna system and that is first described.

3.3.1 Orthogonal Design for a 2 × 1 MISO System: Alamouti STBC

3.3.1.1 Code Description

The first space–time block code was invented by Alamouti. The Alamouti STBC transmits two symbols x_1 and x_2 in two transmission periods as shown in Figure 3.25:

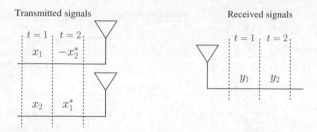

Figure 3.25 Alamouti STBC: transmitted and received signals.

- During the first transmission period, symbol x_1 is transmitted from antenna 1 and symbol x_2 is transmitted from antenna 2.
- During the second transmission period, symbol $-x_2^*$ is transmitted from antenna 1 and symbol x_1^* is transmitted from antenna 2.

Symbols x_1 and x_2 belong to two different codewords (hence are uncorrelated). An equal transmit power $\bar{P}/2$ is assigned to each transmit antenna. Such a transmission has been defined as i.i.d. For a MISO slow fading channel, i.i.d. transmission is only optimal at high SNR, but we have seen in Chapter 2 that it is a reasonable assumption at lower SNRs as well.

The same encoder is used to encode both codewords, but they are encoded independently as becomes apparent below, this is justified because both symbols transmitted through the same channel. The rate of the encoder is determined by the statistics of the channel. The codewords contain the same number of symbols. Pairs of symbols, one symbol from each codeword, are transmitted one after another during two transmission periods.

Another important assumption is that the channel is constant during the two transmission periods over which two STBC symbols are transmitted. This property is essential for the STBC to extract the maximum diversity from the channel.

It is customary to represent a space–time code by a matrix. The Alamouti STBC code matrix is:

$$
\text{Tx antenna} \downarrow \xrightarrow{\text{time}} \begin{bmatrix} x_1 & -x_2^* \\ x_2 & x_1^* \end{bmatrix}. \tag{3.104}
$$

The element (i, j) of a code matrix is the symbol transmitted in time slot i from antenna j. The rate of an STBC is the average number of transmitted symbols per transmission period. The Alamouti STBC transmits two symbols in two transmission periods so the rate of the Alamouti STBC is one. It is said to have *full rate*. Indeed, having a rate higher than one would mean that the same symbol is transmitted more than once from the same antenna, which is useless to obtain full diversity.

We notice that the lines of the STBC matrix in Equation (3.104) are orthogonal. This orthogonality feature gives the Alamouti STBC its full diversity property. We now elaborate further on this point.

3.3.1.2   Orthogonal Transmission

We denote h_i the channel from transmit antenna i to the receive antenna and $\mathbf{h}^T = [h_1 \quad h_2]$ the 1×2 channel vector; y_1 and y_2 are the received signals during the first and second symbol transmission periods and n_1 and n_2 the corresponding additive noise samples. As usual, the noise is assumed Gaussian and i.i.d. The input–output relationship is:

$$\begin{bmatrix} y_1 & y_2 \end{bmatrix} = \mathbf{h}^T \begin{bmatrix} x_1 & -x_2^* \\ x_2 & x_1^* \end{bmatrix} + \begin{bmatrix} n_1 & n_2 \end{bmatrix} \qquad (3.105)$$

To understand how the Alamouti STBC exhibits a full diversity gain, we consider first the noiseless case. The noiseless received signals are:

$$\begin{cases} z_1 = h_1 x_1 + h_2 x_2 \\ z_2 = h_2 x_1^* - h_1 x_2^* \end{cases} \qquad (3.106)$$

In order to easily find the optimal receiver, we determine the equivalent channel corresponding to the transmission of x_1 and x_2. The data content sent during the first and second transmission time is the same (conjugation does not change the information content). A quick trick to make this obvious is to take the conjugate of z_2:

$$\begin{cases} z_1 = h_1 x_1 + h_2 x_2 \\ z_2^* = h_2^* x_1 - h_1^* x_2 \end{cases} \qquad (3.107)$$

It becomes easy to find the equivalent channel corresponding to the transmission of x_1 and x_2. The system becomes equivalent to 2×2 MIMO channel:

$$\begin{bmatrix} z_1 \\ z_2^* \end{bmatrix} = \begin{bmatrix} h_1 & h_2 \\ h_2^* & -h_1^* \end{bmatrix} \begin{bmatrix} x_1 \\ x_2 \end{bmatrix} \qquad (3.108)$$

A remarkable property is that the equivalent channel matrix is orthogonal. This orthogonality is a direct consequence of the orthogonality the Alamouti STBC.

3.3.1.3   Receiver

Let us now come back to the received signals with additive noise, taking the complex conjugate of y_2:

$$\begin{bmatrix} y_1 \\ y_2^* \end{bmatrix} = \begin{bmatrix} h_1 & h_2 \\ h_2^* & -h_1^* \end{bmatrix} \begin{bmatrix} x_1 \\ x_2 \end{bmatrix} + \begin{bmatrix} n_1 \\ n_2^* \end{bmatrix} \tag{3.109}$$

When the MIMO channel matrix is orthogonal, we have seen that the optimal receiver is the matched filter. The two outputs of the matched filter are:

$$\begin{cases} \tilde{y}_{1,\text{MF}} = \left(|h_1|^2 + |h_2|^2 \right) x_1 + \tilde{n}_1 \\ \tilde{y}_{2,\text{MF}} = \left(|h_1|^2 + |h_2|^2 \right) x_2 + \tilde{n}_2 \end{cases} \tag{3.110}$$

where $\tilde{n}_1 = h_1^* n_1 - h_2 n_2^*$ and $\tilde{n}_2 = h_2^* n_1 + h_1 n_2^*$. $\tilde{n}_1$ and $\tilde{n}_2$ are independent and equally distributed as $\mathcal{CN}\left(0, \sigma_n^2[|h_1|^2 + |h_2|^2]\right)$. In the resulting system (3.110), *both symbols x_1 and x_2 see the same channel*. Hence, STBC transmission is equivalent to the generic SISO channel:

$$\tilde{y} = \left(|h_1|^2 + |h_2|^2 \right) x + \tilde{n}, \quad \text{with SNR} = \frac{\bar{P}}{2} \frac{\left(|h_1|^2 + |h_2|^2 \right)}{\sigma_n^2} \tag{3.111}$$

3.3.1.4   Diversity Gain

To easily determine the diversity gain of the Alamouti STBC, we make a parallel with a 1×2 SIMO system. For such a system, the maximal number of independent channels carrying the same symbol is clearly equal to two. The SISO system in Equation (3.111) can also be seen as a two receive antenna SIMO system with channels h_1 and h_2 where the transmit power is reduced to $\bar{P}/2$. We recall that the optimal receiver for this system is the spatial matched filter which leads to the post-processing SNR value in Equation (3.111). We conclude that the maximal diversity gain of the Alamouti STBC is equal to two. Hence the Alamouti STBC achieves a full diversity gain. Another way to determine the diversity gain is to use the definition in Equation 2.82.

Similarly, the system in Equation (3.111) can be seen as a two transmit antenna MISO system where *the CSI is known at the transmitter* and the transmit power is reduced to $\bar{P}/2$. The optimal transmission technique is transmit spatial matched filtering. Again, for such a system, the maximal diversity gain is equal to two. Compared to this system, Alamouti STBC loses by a factor of two in array gain for not knowing the CSI at the transmitter.

3.3.1.5   Alamouti STBC: Minimises the Outage Probability for i.i.d. transmission

STBC transmission is equivalent to the SISO channel in (3.111). Hence, the maximal rate for a reliable communication is equal to $\log_2\left(1 + \dfrac{\bar{P}}{2} \dfrac{\|\mathbf{h}\|^2}{\sigma_n^2} \right)$. It is achieved when the CSI is known at the transmitter. Because the instantaneous value of the CSIT is not known, when transmitting at rate R, the system will be in outage with probability:

$$p_{\text{out}}^{\text{Ala}}(R) = \Pr\left\{ \log_2\left(1 + \frac{\bar{P}}{2}\frac{\|\mathbf{h}\|^2}{\sigma_n^2}\right) < R \right\} \tag{3.112}$$

The Alamouti STBC has the same outage probability as the MISO channel described in Section 2.7.5. Therefore, when the transmission is constrained to be i.i.d., the Alamouti STBC is the optimal code, that is, it minimises the outage probability of the system.

The Alamouti STBC has **full rate** and **full diversity**. Only for two transmit antenna can a space–time block code achieve both properties (except for real valued constellations). STBC designs for more than two transmit antennas can achieve (a) full rate but not full diversity or (b) full diversity but not full rate.

Alamouti STBC transmission is equivalent to a SISO channel with SNR equal to:

$$\text{SNR} = \frac{\bar{P}}{2}\frac{\left(|h_1|^2 + |h_2|^2\right)}{\sigma_n^2} \tag{3.113}$$

Alamouti STBC is optimal in slow fading channels and minimises the outage probability of a 2×1 MISO with CSIT distribution.

3.3.1.6 Multiple Receive Antennas

Alamouti STBC can be used in MIMO communications. It benefits from additional diversity and array gain due to the presence of multiple receive antennas. However, it does not use MIMO multiplexing capabilities. Hence it is suboptimal as it does not achieve the highest possible throughput.

The treatment with multiple receive antennas is very similar to the treatment with a single receive antenna except that we now manipulate vectors. The received signals during the two transmission periods are:

$$\begin{bmatrix} \mathbf{y}_1 \\ \mathbf{y}_2^* \end{bmatrix} = \begin{bmatrix} \mathbf{h}_1 & \mathbf{h}_2 \\ \mathbf{h}_2^* & -\mathbf{h}_1^* \end{bmatrix} \begin{bmatrix} x_1 \\ x_2 \end{bmatrix} + \begin{bmatrix} \mathbf{n}_1 \\ \mathbf{n}_2^* \end{bmatrix} \tag{3.114}$$

$\mathbf{y}_k$ is the $M_R\times 1$ vector of received signals at time k, $k = 1, 2$. $\mathbf{h}_i$ is the $M_R\times 1$ vectorial channel from transmit antenna i, $i = 1, \ldots, M_R$, to the receive antennas. $\mathbf{n}_k$ is the $M_R\times 1$ vector of additive noise samples at time k. The equivalent channel matrix has dimension $M_R\times 2$ and has orthogonal columns.

The optimal receiver for each stream is the matched filter. The equivalent SISO channel for each stream is:

$$\begin{cases} \tilde{y}_{1,\text{MF}} = \left(|\mathbf{h}_1|^2 + |\mathbf{h}_2|^2\right) x_1 + \tilde{n}_1 \\ \tilde{y}_{2,\text{MF}} = \left(|\mathbf{h}_1|^2 + |\mathbf{h}_2|^2\right) x_2 + \tilde{n}_2 \end{cases} \tag{3.115}$$

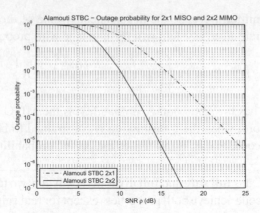

Figure 3.26 Outage probability of the Alamouti STBC for a 2×1 MISO and 2×2 MIMO system and an i.i.d. Rayleigh slow fading channel model (with $E|h_{ij}|^2 = 1$).

$\tilde{n}_1 = \mathbf{h}_1^H \mathbf{n}_1 - \mathbf{h}_2^T \mathbf{n}_2^*$ and $\tilde{n}_2 = \mathbf{h}_2^H \mathbf{n}_1 + \mathbf{h}_1^T \mathbf{n}_2^*$. Finally, the post processing SNR is:

$$\text{SNR} = \frac{\bar{P}}{2} \frac{\left(|\mathbf{h}_1|^2 + |\mathbf{h}_2|^2\right)}{\sigma_n^2} \quad (3.116)$$

In Figure 3.26, we show the outage probability of the Alamouti STBC for a 2×1 MISO and 2×2 MIMO system as a function of the SNR $\rho = \bar{P}/\sigma_n^2$. The transmission rate is equal to four bits per transmission. From the slopes of the curves at high SNR, we can observe that the diversity gain of the Alamouti STBC for a 2×1 MISO is equal to two while the diversity gain of the Alamouti STBC for a 2×2 MIMO system is equal to four. The outage probability is lower for the 2×2 MIMO system thanks to the higher diversity gain and the array gain.

3.3.2 STBC for More than Two Transmit Antennas

We first generalise definitions introduced during the description of the Alamouti STBC:

- A space–time code is represented using the following matricial form:

$$
\begin{array}{c}
\text{time} \\
\longrightarrow
\end{array}
$$

$$
\text{Tx antenna} \downarrow
\begin{bmatrix}
x_{11} & x_{12} & \cdots & x_{1T} \\
x_{21} & x_{22} & \cdots & x_{2T} \\
\vdots & \vdots & \vdots & \vdots \\
x_{M_T 1} & x_{22} & \cdots & x_{2T}
\end{bmatrix}
\quad (3.117)
$$

x_{ij} is the symbol transmitted in time slot i from antenna j. Row i describes the transmission from transmit antenna i at all symbol transmission time. Column j describes the transmission from all transmit antenna at symbol transmission time j.

- The code rate of an STBC is the number of symbols transmitted on average over a block.
- If the rate is equal to 1, then the STBC has *full rate*.
- With M_T transmit antennas (and 1 receive antenna), the maximal diversity gain is M_T. A STBC is said to have *full diversity* gain if its diversity gain is equal to M_T.

The data to be transmitted is encoded, using the same encoder, into multiple codewords, or blocks, of same duration. Multiple copies of the same block are transmitted in space and in time. The STBC spreads over a number of T block transmissions and over all transmit antennas. Hence, decoding is based jointly on T blocks at the receiver. The main assumptions associated with STBC transmission are as follows.

Main assumptions for STBC

▶ MISO channel coefficients are known at the receiver.
▶ Slow fading channel: instantaneous CSIT not known at the transmitter, but its statistics are known.
▶ Channel is constant over the duration of the STBC (transmission of symbol block).
▶ Unlike Alamouti STBC, the power is not always equally distributed across the transmit antennas (unequal power allocation might be necessary to guarantee orthogonality of the STBC).
▶ Noise at receiver is ZMCCS, temporally and spatially white: $n \sim \mathcal{CN}(0, \sigma_n^2)$.

3.3.2.1 Orthogonal and Quasi-orthogonal Designs

For more than two transmit antennas, two classes of STBC codes can be distinguished: the class of orthogonal STBC and the class of quasi-orthogonal STBC.

- **Orthogonal STBC**: The lines of the STBC matrix are orthogonal. The advantage of orthogonal STBC is twofold: (a) they have full diversity and (b) the optimal receiver is very simple as it is a simple matched filtering. The disadvantage is that those codes do not achieve full rate, with noticeable exception of the Alamouti STBC for two transmit antennas (as well as real valued constellations).
- **Quasi-orthogonal STBC**: The lines of the STBC matrix are not orthogonal. The orthogonality is sacrificed for rates that are higher than the orthogonal counterpart. However, the optimal receiver is more complex (ML receiver in general).

Designs for orthogonal and quasi orthogonal codes can easily be found in the literature. In the following paragraph, we give the example of STBC designed for a four-transmit antenna system.

3.3.2.2   Orthogonal Design: Number of tx Antennas Larger than Two

For a real valued constellations, full rate and full diversity STBC can be designed. For a four-transmit antenna system, such a STBC is:

$$C_1^r = \begin{bmatrix} x_1 & -x_2 & -x_3 & -x_4 \\ x_2 & x_1 & x_4 & -x_3 \\ x_3 & -x_4 & x_1 & x_2 \\ x_4 & x_3 & -x_2 & x_1 \end{bmatrix} \tag{3.118}$$

When the constellation is complex, an orthogonal STBC cannot have full rate. For a four-transmit antenna system, an STBC with rate 1/2 is:

$$C_{1/2} = \begin{bmatrix} x_1 & -x_2 & -x_3 & -x_4 & x_1^* & -x_2^* & -x_3^* & -x_4^* \\ x_2 & x_1 & x_4 & -x_3 & x_2^* & x_1^* & x_4^* & -x_3^* \\ x_3 & -x_4 & x_1 & x_2 & x_3^* & -x_4^* & x_1^* & x_2^* \\ x_4 & x_3 & -x_2 & x_1 & x_4^* & x_3^* & -x_2^* & x_1^* \end{bmatrix} \tag{3.119}$$

Orthogonal codes with higher rates can be designed. For example, the following orthogonal code has rate 3/4. For this code, the transmit power is not constant across antennas, which is required to have orthogonality.

$$C_{3/4} = \begin{bmatrix} x_1 & -x_2^* & \frac{x_3^*}{\sqrt{2}} & \frac{x_3^*}{\sqrt{2}} \\ x_2 & x_1^* & \frac{x_3^*}{\sqrt{2}} & -\frac{x_3^*}{\sqrt{2}} \\ \frac{x_3}{\sqrt{2}} & \frac{x_3}{\sqrt{2}} & \frac{-x_1-x_1^*+x_2-x_2^*}{2} & -\frac{\left(x_1-x_1^*-x_2-x_2^*\right)^*}{2} \\ \frac{x_3}{\sqrt{2}} & -\frac{x_3}{\sqrt{2}} & \frac{x_1-x_1^*-x_2-x_2^*}{2} & \frac{\left(-x_1-x_1^*+x_2-x_2^*\right)^*}{2} \end{bmatrix} \tag{3.120}$$

3.3.2.3   Quasi-Orthogonal Design: Number of tx Antennas Larger than Two

A quasi-orthogonal STBC (QOSTBC) for four-transmit antennas is as follows:

$$C_1^{\text{QOSTBC}} = \begin{bmatrix} x_1 & -x_2^* & -x_3^* & x_4 \\ x_2 & x_1^* & -x_4^* & -x_3 \\ x_3 & -x_4^* & x_1^* & -x_2 \\ x_4 & x_3^* & x_2^* & x_1 \end{bmatrix} \tag{3.121}$$

This STBC has full rate but only partial orthogonality and hence does not achieve full diversity.

This partial orthogonality is confirmed by observing the equivalent channel corresponding to the transmission of x_1, x_2, x_3 and x_4. Based on the same trick we used for the Alamouti STBC (see Equation (3.106)), this equivalent channel is:

$$H = \begin{bmatrix} h_1 & h_2 & h_3 & h_4 \\ h_2^* & -h_1^* & h_4^* & -h_3^* \\ h_3^* & h_4^* & -h_1^* & -h_2^* \\ h_4 & -h_3 & -h_2 & h_1 \end{bmatrix} \tag{3.122}$$

We can observe that not all column pairs are orthogonal.

3.3.2.4   Comparison between Orthogonal STBC and Quasi-orthogonal STBC

Figure 3.27 shows the symbol error rate (SER) at the output of the ZF receiver as a function of the SNR $\rho = \bar{P}/\sigma_n^2$ for the orthogonal STBC of Equation (3.119) and that of the quasi-orthogonal STBC of Equation (3.121), assuming that a QPSK constellation is transmitted over an i.i.d. complex Gaussian (Rayleigh) fading channels with $E|h_{ij}|^2 = 1$. The SER for a SISO channel is shown as a reference. Two main observations can be drawn from the figure:

- We have defined the diversity gain as the slope of the outage probability curves at high SNR. For an uncoded signal, the diversity gain is also the slope of the SER (or BER) curve at high SNR. Note that this last statement is not always true for a coded input. While OSTBC achieves a full diversity gain, QOSTBC does not.

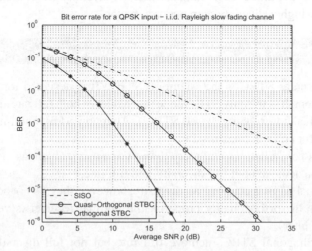

Figure 3.27   SER of OSTBC and QOSTBC.

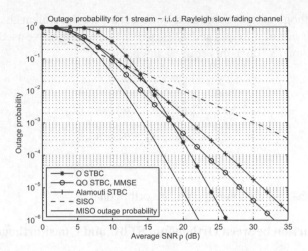

Figure 3.28 Outage probability of OSTBC and QOSTBC for a selected rate R=2 bits per transmission and an i.i.d. Rayleigh slow fading channel.

- QOSTBC has a worse SER than OSTBC. For a fixed input constellation, the SER of QOSTBC is degraded due to the inter-stream interference (or nonorthogonality of the QOSTBC). However, the OSTBC shown has half the rate of the QOSTBC.

In practice, for enhanced performance, the transmission rate should be adapted to the channel quality, for which the outage probability becomes relevant. Figure 3.28 shows the outage probability of a given output of the ZF or MMSE receiver (all the outputs have the same statistics). In the low SNR range, the schemes with lowest diversity perform better. However, at the higher SNR range, when the diversity effect kicks in, the designs benefiting from a higher diversity gain become better.

Space–Time Block Codes

▶ The most popular STBC is the Alamouti STBC. It is designed for a two-transmit antenna system. It is the only STBC code that achieves both full rate and full diversity (except for real constellation based STBC). Alamouti STBC minimises the outage probability for an i.i.d. transmission.

▶ For complex constellations and more than two transmit antennas, no STBC can be designed that achieve both full diversity and full rate.
  - Orthogonal STBCs achieve full diversity, however, they do not achieve full rate: thanks to the orthogonality property, the optimal receiver is very simple (i.e. matched filter).
  - Quasi-orthogonal STBCs achieve full rate but not full diversity: the optimal receiver is more complex.

3.4 D-Blast

The Diagonal Bell Labs Layered Space Time (D-Blast) transceiver is a *spatial multiplexing* architecture aiming at the highest possible diversity gain, while still achieving a high rate. Like STBC, it is targeted for slow fading channels with knowledge of the statistics of the CSIT, but not its instantaneous value. However, unlike the STBCs described in Section 3.3, it exhibits spatial multiplexing capabilities. It is the optimal transceiver when the channel is slow fading and under asymptotic conditions (large number of transmitted codewords) that are detailed later on.

Main assumptions for D-Blast

▶ MIMO channel coefficients are known at the receiver.
▶ Slow fading channel: instantaneous CSIT not known at the transmitter, but its statistics are known.
▶ Channel is constant over the duration of a transmission (see Figure 3.29).
▶ Same power is allocated to all streams.
▶ Noise at receiver is ZMCCS, temporally and spatially white: $\mathbf{n} \sim \mathcal{CN}(0, \sigma_n^2 \mathbf{I})$.

D-Blast is based on the following two features:

1. **Diagonal Encoding:** Codewords or layers are spread diagonally across antennas.
2. **Diagonal Decoding:** Layers are decoded diagonally and removed one by one.

Below, D-Blast is mostly described for a two transmit antenna system. The generalisation to larger systems is straightforward.

3.4.1 Diagonal Encoding

The encoding is illustrated in Figure 3.29. Each codeword is divided into M_T parts of same length. Each codeword forms a diagonal layer. For the 2×2 example, the ith transmitted codeword is divided into two parts as:

$$X_i = \{X_i^{(\text{Part 1})}, X_i^{(\text{Part 2})}\}. \tag{3.123}$$

Each codeword has the same constant rate determined to comply with a targeted outage probability. The different parts of each codeword are transmitted successively from a different antenna:

• First transmission time: $X_1^{(\text{Part 1})}$ is transmitted from the first antenna while the other antenna remains idle.

- Second transmission time: $X_1^{(\text{Part 2})}$ is transmitted from antenna 2 and the first part of the second codeword $X_2^{(\text{Part 1})}$ is transmitted from antenna 2.
- Encoding is done similarly for the subsequent codewords until the last transmission time where the second part of the last codeword is transmitted from antenna 2 while the other antenna remains idle.

3.4.2 Diagonal Decoding

The general principle of the decoding follows the principle of successive interference cancellation as follows:

- A receiver is applied to get a soft estimate of each part of the same codeword.
- The information about each part of a codeword is combined to perform decoding.
- Once the decoding is achieved, the contribution of the codeword is reconstructed and removed. Decoding of the next codeword proceeds similarly.

In Figure 3.29, we detail how we get a soft estimate of each part of a codeword using a MMSE receiver.

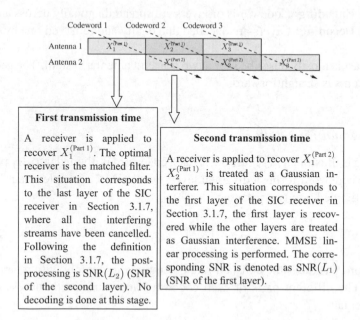

Figure 3.29   D-Blast for a two-transmit antenna system. Steps to get soft estimates of each part of a codeword.

3.4.3   D-Blast: Outage Optimal

For a given channel realisation, each part of a codeword sees a channel with SNR equal to $\text{SNR}(L_k)$. The key observation is that $\text{SNR}(L_k)$ is also the SNR of each layer of the MMSE-SIC receiver as described in Section 3.1.7. For each part of a codeword, the maximal achievable rate for reliable communication is $\log_2 [1 + \text{SNR}(L_k)]$. Hence, the highest rate achievable for a whole codeword is $\sum_{k=1}^{M_T} \log_2 [1 + \text{SNR}(L_k)]$.

In Section 3.1.7, we have seen that it is equal to maximal achievable rate for an i.i.d. transmission

$$\mathcal{R}_{\max} = \log_2 \det \left(1 + \frac{\bar{P}}{M_T \sigma_n^2} HH^H \right) \tag{3.124}$$

D-Blast achieves this rate only asymptotically because D-Blast suffers from a rate loss at the beginning of the transmission to initialise the decoder as well as at the end of transmission. However, we will assume that the number of codewords to transmit is large so that this loss becomes negligible.

If the channel were known at the transmitter, we know that there exist a code achieving this rate. As the transmitter does not know the channel but its statistics, it can adjust its transmission rate to comply with a targeted outage probability based on the channel distribution. The outage probability of D-Blast for selected transmit rate R is:

$$P_{\text{out}} = \Pr \left\{ \log_2 \det \left(1 + \frac{\bar{P}}{M_T \sigma_n^2} HH^H \right) < R \right\} \tag{3.125}$$

This is also the outage probability of the MIMO slow fading channel (with i.i.d. transmission). So we conclude that D-Blast asymptotically minimises the outage probability for a MIMO slow fading channel with i.i.d. transmission.

3.4.4   Performance Gains

- **Multiplexing gain.** D-Blast exploits up to $\min(M_T, M_R)$ spatial degrees of freedom (in the soft estimation of the last part of a codeword.)
- **Antenna/power gain.** D-Blast benefits from an array gain that has a maximal value of M_R when the first part of the codeword is estimated.
- **Diversity gain.** D-Blast achieves the maximal diversity gain $M_T M_R$. The maximal diversity gain is reached with MMSE-SIC at the receiver. One has to be careful when considering the performance of MMSE-SIC applied to D-BLAST at high SNR ρ. They are not equal to the performance of ZF-MMSE, although the post-processing $\text{SNR}(L_k)$ of each individual layer of MMSE-SIC tends to the SNR obtained by the ZF-SIC. The reason is that the layers of MMSE-SIC are correlated. Coding is done across layers and manages to favourably exploit this correlation, bringing the diversity gain that goes missing in ZF-SIC.

3.4.5   Error Propagation

D-Blast looks advantageous especially for its full diversity gain. However, it suffers from a significant flaw: *error propagation*. For a slow fading channel, we can only guarantee a fixed value for the outage probability. Errors might occur and might propagate to the next layers when, after decoding, a diagonal codeword is removed from the received signal. The outage probability given in Equation (3.125) does not account for error propagation. Note that in V-Blast also uses successive interference cancellation. However, for fast fading channels, coding is done over many channel fades thus guaranteeing a negligible error probability for each layer and hence avoiding error propagation.

3.4.6   Numerical Evaluations: Comparison of D-Blast with STBC

Figure 3.30 shows the performance of D-Blast compared to some of the STBC schemes seen in Section 3.3, for a 4x4 i.i.d. Rayleigh fading MIMO channel. We recall that the STBC schemes seen in section 3.3 are designed for a MISO channel and benefit from transmit diversity. STBC schemes are suboptimal for a MIMO system as they do not exploit MIMO multiplexing capabilities, rather they benefit from receive antenna gain and diversity but not from multiplexing. Figure 3.30 shows the outage probability of the following schemes for an input rate of 2 bits per transmission:

- D-Blast: Asymptotic performance when neglecting the loss due to transmission initialisation and termination as well as error propagation.

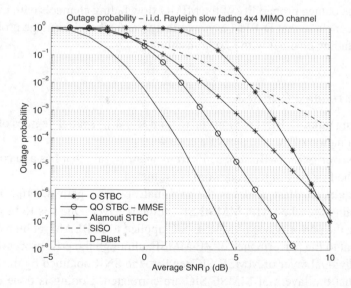

Figure 3.30   Outage probability of OSTBC and QOSTBC for a selected rate R=2 bits per transmission and an i.i.d. Rayleigh slow fading channel.

- Orthogonal STBC as in (3.119).
- Quasi-orthogonal STBC as in (3.121) where a MMSE receiver is applied.
- Alamouti STBC, where two of the transmit antennas are selected for transmission.
- SISO channel.

We observe that the performance of D-Blast is indeed much better than the STBC schemes. Furthermore, for a targeted outage probability above 10^{-8}, QOSTBC outperforms OSTBC as well as Alamouti STBC.

D-Blast
▶ D-Blast is a spatial multiplexing structure providing both a high throughput and high diversity gain. It targets slow fading MIMO channels with CSIT distribution. It minimises the outage probability (asymptotically in the number of transmitted codewords) of a MIMO slow fading channel when the transmission is constrained to be i.i.d.
▶ D-Blast is based on a diagonal encoding where codewords are divided into M_T parts and transmitted diagonally from each antenna (see Figure 3.29). Likewise, the decoding is performed diagonally based on the principle of successive interference cancellation.

3.5   Chapter Summary Tables

The main results on MIMO transceivers contained in this chapter are presented in the form of tables for easy reference. Three classes of transceivers are described.

▶ **Receivers for spatial multiplexing architectures**
    — Matched filter
    — Zero-forcing receiver
    — MMSE receiver
    — Successive interference cancellation: MMSE-SIC and V-Blast
▶ **Transmit and receive beamforming with CSIT and CSIR**
    — Single beam techniques:
        ∗ Transmit matched filter
        ∗ Receive matched filter
        ∗ Maximal eigenmode transmission
    — Multiple beam technique (spatial multiplexing architecture):
        ∗ Eigenmode transmission with waterfilling power allocation
▶ **Space–time architectures for slow fading channels**
    — Single stream (MISO):
        ∗ Space–time block codes
    — Multiple streams (spatial multiplexing architecture):
        ∗ D-Blast

MIMO receivers: expressions and post-processing SNRs

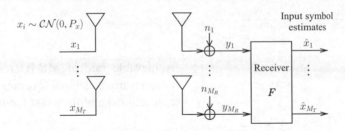

		Receiver	Post-processing SNR
MF	Block	$F_{\text{MF}} = \mathcal{D}_{\text{MF}}^{-1} H^{\text{H}}$	$\times$
ISI modelled as white noise	Stream	$F_{\text{ZF}}(k) = \dfrac{\mathbf{h}_k^{\text{H}}}{\|\mathbf{h}_k\|^2}$	$\text{SNR}_{\text{MF}}(k) =$ $\dfrac{P_x\|\mathbf{h}_k\|^2}{\sum_{i \neq k} |\mathbf{h}_k^{\text{H}}\mathbf{h}_i|^2/\|\mathbf{h}_k\|^2 + \sigma_n^2}$
ZF	Block	$F_{\text{ZF}} = K_{\text{ZF}}^{-1} H^{\text{H}}$	$\text{SNR}_{\text{ZF}}(k) = \dfrac{P_x}{\sigma_n^2 [K_{\text{ZF}}^{-1}]_{kk}}$
ISI modelled as deterministic ISI elimination	Stream	$F_{\text{ZF}}(k) = \dfrac{\mathbf{h}_k^{\text{H}} P_{\bar{H}_k}^{\perp}}{\mathbf{h}_k^{\text{H}} P_{\bar{H}_k}^{\perp} \mathbf{h}_k}$	$\text{SNR}_{\text{ZF}}(k) = \dfrac{P_x}{\sigma_n^2} \mathbf{h}_k^{\text{H}} P_{\bar{H}_k}^{\perp} \mathbf{h}_k$
UMMSE	Block	$F_{\text{UMMSE}} = \mathcal{D}_{\text{MMSE}}^{-1} K_{\text{MMSE}}^{-1} H^{\text{H}}$	$\text{SNR}_{\text{UMMSE}}(k) = \dfrac{P_x}{\sigma_n^2 [K_{\text{MMSE}}^{-1}]_{kk}}$
ISI modelled as coloured noise	Stream	$F_{\text{UMMSE}}(k) = \dfrac{\mathbf{h}_k^{\text{H}} \bar{R}_k^{-1}}{\mathbf{h}_k^{\text{H}} \bar{R}_k^{-1} \mathbf{h}_k}$	$\text{SNR}_{\text{UMMSE}}(k) = P_x \mathbf{h}_k^{\text{H}} \bar{R}_k^{-1} \mathbf{h}_k$
MMSE-SIC *ISI modelled as coloured noise*	Stream	$F_{\text{UMMSE}}(k) = \dfrac{\mathbf{h}_k^{\text{H}} \bar{R}_{L_k}^{-1}}{\mathbf{h}_k^{\text{H}} \bar{R}_{L_k}^{-1} \mathbf{h}_k}$	$\text{SNR}_{\text{UMMSE}}(k) = P_x \mathbf{h}_k^{\text{H}} \bar{R}_{L_k}^{-1} \mathbf{h}_k$

Notations

MF:

$$\mathcal{D}_{\text{MF}} = \text{diag}\left[H^{\text{H}} H \right]$$

$$K_{\text{ZF}} = H^{\text{H}} H$$

$$P_{\bar{H}_k}^{\perp} = I - \bar{H}_k \left(\bar{H}_k^{\text{H}} \bar{H}_k \right)^{-1} \bar{H}_k^{\text{H}},$$

$$\bar{H}_k = H \text{ without its } k^{th} \text{ column}$$

UMMSE:

$$K_{\text{MMSE}} = H^{\text{H}} H + \frac{\sigma_n^2}{P_x} I$$

$$\bar{R}_k = P_x \sum_{i \neq k} \mathbf{h}_i \mathbf{h}_i^{\text{H}} + \sigma_n^2 I$$

$$\mathcal{D}_{\text{MMSE}} = \text{diag}\left[K_{\text{MMSE}}^{-1} H^{\text{H}} H \right]$$

MMSE-SIC:

$$\bar{R}_{L_k} = P_x \sum_{i > k} \mathbf{h}_i \mathbf{h}_i^{\text{H}} + \sigma_n^2 I$$

MIMO receivers: performance at low and high SNR

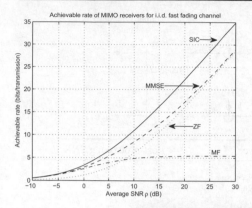

Achievable rate of MIMO receivers for a time-invariant channel (average over channel realisations) and for a fast fading channel.

Time-invariant and fast fading channels		
	Low SNR	High SNR
MF		ISI limited
ZF	Worse than MF	
MMSE	Equivalent to MF	Equivalent to ZF
MMSE-SIC	Equivalent to MF-SIC (with reduced ISI at each step)	Equivalent to ZF-SIC (with reduced ISI at each step)
	Achieves maximal rate for fixed channel realisation. Achieves capacity for fast fading channel	

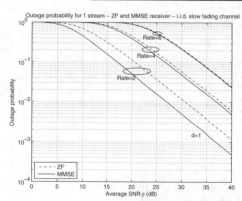

Outage probability of ZF and MMSE receivers for a slow fading channel for rate per stream equal to 2, 4 and 6 bits/transmission. ZF and MMSE are not equivalent.

Diversity gain per stream: slow fading channel	
	Diversity gain
ZF/MMSE	$M_R - M_T + 1$
MMSE-SIC (at step k)	$M_R - M_T + k - 1$

Transmit and receive beamforming with CSIT and CSIR

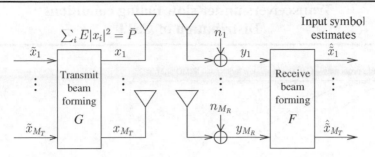

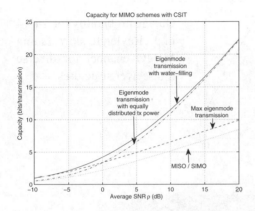

Achievable rate of beamforming schemes averaged over channel realisations drawn from a i.i.d. Rayleigh distribution.

		Transmit/receive filter	Post-processing SNR
Tx MF	Single stream	$G_{\mathrm{TMF}} = \mathbf{h}^{\mathrm{H}}/\|\mathbf{h}\|$ $(F_{\mathrm{TMF}} = 1/\|\mathbf{h}\|)$	$\mathrm{SNR}_{\mathrm{TMF/RMF}} = \dfrac{\bar{P}\|\mathbf{h}\|^2}{\sigma_n^2}$
Rx MF	Single stream	$F_{\mathrm{RMF}} = \mathbf{h}^{\mathrm{H}}/\|\mathbf{h}\|^2$ $(G_{\mathrm{RMF}} = 1)$	
Max eigenmode	Single stream	$G_{\mathrm{MEig}} = \bar{P}\mathbf{v}_{\max}$	$\mathrm{SNR}_{\mathrm{MEig}} = \dfrac{\bar{P}\lambda_{\max}^2}{\sigma_n^2}$
Eigenmode	Multi-stream	$G_{\mathrm{Eig}} = V\mathbf{P}_{\mathrm{root}}^{o\,1/2}$ $F_{\mathrm{Eig}} = U^{\mathrm{H}}$	$\mathrm{SNR}_{\mathrm{Eig}}(k) = \dfrac{P_k^o \lambda_k^2}{\sigma_n^2}$

Notations

SVD of H	$H = U\Lambda V^{\mathrm{H}}$
	eigenvalue k of H
$V_{\max}$	right eigenvector associated with the largest eigenvalue $\lambda_{\max}$
$\mathbf{P}_{\mathrm{root}}{}^o$	$\mathrm{diag}[P_1^o \ldots P_{r_H}^o\, 0 \ldots 0]$, $M_T \times M_T$ diagonal matrix with waterfilling power allocation, $\mathbf{P}_{\mathrm{root}^o}^{1/2} = \mathrm{diag}[P_1^{o\,1/2} \ldots P_{r_H}^{o\,1/2}\, 0 \ldots 0]$

Transceivers under slow fading conditions
Distribution of CSIT

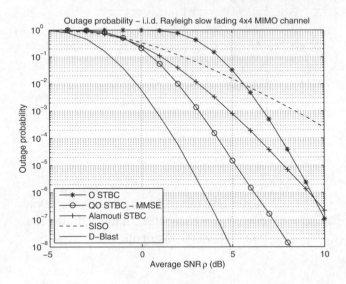

Outage probability for a 4×4 i.i.d. Rayleigh slow fading MIMO channel for diverse transceiver structures.

Space–time block codes

Orthogonal STBC	Quasi-orthogonal STBC
Lines of STBC matrix orthogonal achieves full diversity not full rate simple optimal receiver = matched filter	Lines of STBC matrix quasi-orthogonal does not achieve full diversity higher rate than orthogonal counterpart optimal receiver is more complex = ML

Alamouti STBC:
Optimal scheme for a 2×1 slow fading MISO
channel (minimises outage probability).
Full rate and full diversity.

D-Blast

Optimal schemes for a MIMO slow fading channel
(minimises outage probability).
Achieves full diversity $M_T M_R$

3.6 Further Reading

The topics treated in this chapter require basic knowledge of linear algebra (e.g. as in (Strang 2009)) and probability (e.g. Papoulis (1965)).

The D-Blast and V-Blast algorithms have been presented and evaluated in Foschini (1996), Foschini and Gans (1998), Golden *et al.* (1999), Wolniansky *et al.* (1998).

Several books present space–time codes in detail such as Larsson and Stoica (2003) and Paulraj *et al.* (2003). The trade-off between diversity and multiplexing was analysed in Tse *et al.* (2004) and several codes have been developed at different operational points of the trade-off curve. The topic of transmit antenna diversity was studied, among others, in Tarrokh *et al.* (1999), where the determinant criterion was analysed. The delay-diversity scheme was introduced in Seshadri *et al.* (1997). The Alamouti scheme was introduced in Alamouti (1998) and generalised to orthogonal designs in Tarrokh *et al.* (1999).

4

MIMO Channel Models

Tim Brown and Persefoni Kyritsi

SISO systems deal with the link between a single transmit antenna and a single receive antenna. In the channel models for the first generation wireless systems, the parameter of interest was the received signal power: its average value, how it varies as a mobile station moves within a small area, how it changes when the mobile station is at different distances from the base station. Therefore these systems concentrated on the characterisation of the distance dependence of the received power, as well as its first and second order statistical properties over small and wide areas.

As wireless systems evolved to accommodate higher data rates, the bandwidth of mobile systems increased and the temporal dispersion of the received signal became noticeable: the shorter pulse duration[1] made it easier to distinguish between individual paths (or rays) of different lengths in a radio environment that the signal had followed between the transmitter and the receiver. Therefore it became important to also include the temporal dispersion among the features necessary to describe the wireless channels. In this chapter, the key aspects of SISO channels and the corresponding models, for both narrow and wide system bandwidths, are shown.

MIMO channels deal with the links between several transmit antennas and several receive antennas. Therefore they include several SISO channels which, depending on the environment, will have different interdependence relations with each other. Modelling MIMO channels is not simply a case of creating multiple SISO channels independent of each other, but rather of finding ways to capture and model their individual and joint variation. As shown in Chapters 2 and 3, the interdependence of the MIMO channels impacts

---

[1]Bandwidth is inversely proportional to symbol time, thus short pulse duration has high bandwidth.

Practical Guide to the MIMO Radio Channel with MATLAB® Examples, First Edition.
Tim Brown, Elisabeth De Carvalho and Persefoni Kyritsi.
© 2012 John Wiley & Sons, Ltd. Published 2012 by John Wiley & Sons, Ltd.

the performance of the communication system in terms of diversity gain, beamforming, multiplexing gain and capacity.

In light of this, the best known and most widely used MIMO channel models are presented in this chapter. The models considered are classified as deterministic (e.g. ray-tracing) or stochastic (e.g. correlation) based, and differ in complexity and accuracy, which makes them therefore more suitable or less suitable for different applications.

4.1   SISO Models and Channel Fundamentals

In Chapter 2 we saw how the received signal $y(t)$ is related to the transmitted signal $x(t)$:

$$y(t) = h(t)x(t) + n(t) \tag{4.1}$$

$h(t)$ determines the channel coefficient, that is, the multiplicative term that decreases or increases the input $x(t)$. The additive term $n(t)$ incorporates background thermal noise as well as other potential sources of noise and also radio interference from other users of the same frequency that are in the vicinity. Channel modelling focuses on $h(t)$, and first we will look at how we model $h(t)$ for a SISO wireless channel . At this stage we will consider single frequency situations, that is, *narrowband channels*.

4.1.1   Models for the Prediction of the Power

We consider $h(t)$ as a random quantity because it has different values for different points in space, or equivalently it is a random quantity over time as a receiver is moving in space or when the objects in the environment between the transmitter and the receiver are moving. It is then interesting to characterise the statistics of this variation.

From Equation (4.1), the ratio of the power of the signal $y(t)$ received by an antenna (under the idealised assumption of no noise), P_{Rx} compared to the power of the signal $x(t)$ transmitted by an antenna, P_{Tx} defines the channel coefficient as a function of the channel coefficient:[2]

$$\frac{P_{Rx}}{P_{Tx}} = |h(t)|^2 \tag{4.2}$$

To determine the channel coefficient at a given distance r from the transmitter while the receiver is moving at a given velocity, we distinguish three scales of variation or fading: the *small scale fading*, h_{small} otherwise known as multipath fading, the *large scale fading*, h_{LARGE}, which is also known as shadowing, and the *path loss*, h_{pl}.

---

[2] The minimum value of P_{Rx} defines the receiver sensitivity, that is, a receiver has a lower limit as to how much power it can receive.

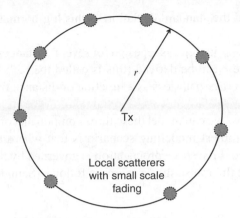

Figure 4.1   Fading components of a mobile radio channel.

The path loss is a deterministic effect and depends on the distance between the transmitter and the receiver, whereas large and small scale fading are random. The observed channel coefficient is the composite result of the superposition of these three different factors:

$$h(t) = h_{\mathrm{pl}} h_{\mathrm{LARGE}} h_{\mathrm{small}}(t) \tag{4.3}$$

The role of these three components is explained in detail with reference to Figure 4.1:

1. **Small scale fading:** Let us assume that we take a point that is at a distance r from the transmitter (Tx) and we take local measurements of the channel coefficient around this point. These would correspond to measurements in the small area of a shaded circle in Figure 4.1. All the points in this circle are at approximately the same distance r from the transmitter, Tx. Within this small area, the channel coefficient varies according to what is known as small scale fading. The details of the statistical behaviour will be discussed later, but for now the parameter of interest is the mean of the channel gain σ_m^2. A different mean value can be calculated for each one of the small circles. Since we are only considering movement of a receiver within one of these circles, only $h_{\mathrm{small}}(t)$ is considered to be time variant.
2. **Large scale fading (shadowing):** Clearly, although all the shaded circles in Figure 4.1 are at the same distance from the transmitter, not all the shaded circles will have the same mean signal power. This phenomenon is called large scale fading or shadowing. For example, if there are obstacles (e.g. large buildings) between the transmitter and some shaded circles, these shaded circles would be expected to have a lower average channel coefficient.

    Therefore σ_m^2 is a random quantity and the distribution of all the σ_m^2 becomes relevant. It has been experimentally observed that the distribution of the σ_m^2 can be very well approximated by a log-normal distribution. Again the details will be presented later,

and let the mean and the standard deviation of this log-normal distribution be μ_s and σ_s respectively.
3. **Path loss:** If the process in step 2 is repeated for several distances r, then the dependence of μ_s on the distance r can be derived (this is called the path loss law), and that will help calculate the average path-loss as a function of distance from the transmitter.

Let us now look at how we can model these three components of the channel coefficient mathematically. The simplest modelling scenario is that where the transmitter and the receiver are in free space. Free space propagation is governed by the Friis equation which relates the received and the transmitted powers as follows (Saunders and Aragon-Zavala 2007):

$$\frac{P_{Rx}}{P_{Tx}} = |h_{pl,fs}(r)|^2 = G_{Tx}G_{Rx}\left(\frac{\lambda}{4\pi r}\right)^2 \tag{4.4}$$

$\lambda = \frac{c}{f}$ is the wavelength f is the frequency and c stands for the speed of light. G_{Tx} and G_{Rx} are the antenna gains at the transmit and the receive end and r is the distance between transmitter and receiver. This equation is only suitable when there is free space and no ground or other major reflective objects between the transmitter and receiver.

A more appropriate model for typical circumstances is to assume that the transmitter and receiver antennas are placed above a perfectly conducting ground plane at heights l_{Tx} and l_{Rx} respectively. At large enough distances, the channel coefficient becomes independent of wavelength at suitably low frequencies if $l_{Tx}, l_{Rx} << r$ and is given as a rough approximation by the following equation (Saunders and Aragon-Zavala 2007):

$$|h_{pl,pg}(r)|^2 = G_{Tx}G_{Rx}\left(\frac{l_{Tx}^2 l_{Rx}^2}{r^4}\right) \tag{4.5}$$

It is interesting to observe that in Equation (4.4) the gain decays as $1/r^2$, whereas in Equation (4.5) the power decays proportionally to $1/r^4$, that is, the reflection from the ground causes an increase in path loss exponent.[3]

In actual environments, neither the free-space model nor the perfect ground plane model reflect the true picture. In such cases, the channel is assumed to follow the free space model up to a distance r_{break} and then decay with a different path loss exponent γ (i.e. proportionally to $r^{-\gamma}$) for distances greater than r_{break}. The path loss is then calculated by an equation of the form (Jakes 1974):

$$|h_{pl}(r)|^2 = \begin{cases} G_{Tx}G_{Rx}\left(\frac{\lambda}{4\pi r_{break}}\right)^2 \left(\frac{r_{break}}{r}\right)^\gamma, & r \geq r_{break} \\ G_{Tx}G_{Rx}\left(\frac{\lambda}{4\pi r}\right)^2, & r \leq r_{break} \end{cases} \tag{4.6}$$

---

[3] The path loss exponent is the exponent to which $1/r$ is raised. In the free space case, the path loss exponent is 2, whereas in the situation above perfect ground, the path loss exponent is 4.

The path loss exponent γ has been derived from fits to actual measured data and typically has values between 2 and 4. Clearly, if the path loss exponent is 2, then the condition of free space is met and the equation simplifies to Friis equation.

$h_{pl}(r)$ describes the channel coefficient's deterministic dependence on distance, averaged over several locations. When there are large objects such as buildings between the transmitter and receiver, they introduce a shadowing loss, which is captured by the term h_{LARGE} in Equation (4.3). It depends on the size, design and materials that the objects are made of, and the resultant losses are modelled statistically by a log-normal distribution, that is, the shadowing taken in logarithmic scale (dB) follows a normal (Gaussian) distribution.

Let $h_{LARGE,dB}$ be the large scale fading h_{LARGE} expressed in decibels (dB), that is, $h_{LARGE,dB} = 10\log_{10}(|h_{LARGE}|^2)$.[4] It follows the normal (Gaussian) distribution the probability density function is:

$$Pr\{h_{LARGE,dB}\} = \frac{1}{\sqrt{2\pi\sigma_s^2}}e^{\frac{-|h_{LARGE,dB}|^2}{2\sigma_s^2}} \qquad (4.7)$$

σ_s is a standard deviation in decibels and it typically 4–8 dB (again the value has been experimentally determined by fits to measured channels).

Path loss (h_{pl}) is a deterministic quantity (i.e. it is determined by physical input parameters like the distance r, the path loss exponent γ etc.). Large scale fading (h_{LARGE}) instead is a random quantity. Their combination ($|h_{pl}|^2|h_{LARGE}|^2$) is a log-normal distributed random variable with mean $10\log_{10}(|h_{pl}|^2)$, because $10\log_{10}(|h_{pl}|^2|h_{LARGE}|^2) = 10\log_{10}(|h_{pl}|^2) + h_{LARGE,dB}$. The combination of the deterministic path loss component and the random large scale fading determine the mean power observed over a local area.

Let us now concentrate on the reasons behind the fluctuations of the signal over a small area. The transmitting antenna in a wireless system sends out an electromagnetic signal around a carrier frequency f_c. The signal can follow many different routes to reach the intended receiver. Along these routes, it can be reflected off various surfaces (buildings etc.), diffracted around corners or scattered off various objects (e.g. foliage). The signal components that follow these different paths are called multipath components and have different amplitudes and phases, and arrive at the receiver from different angles. At the receiver side, they add coherently (in phase) at some locations or incoherently (out of phase) at others, leading to a variation of the received power over space. This phenomenon is known as 'small scale' fading because these variations occur over a local area where they are due to the same actual multipath components.[5] The random change in multipath

---

[4] It is easy to convert $h_{LARGE,dB}$ back to linear scale to obtain the shadowing.
[5] Keep in mind that the phase variation of a plane wave over $[0, 2\pi]$ occurs over distances of one wavelength. Thinking of current systems that operate around 2 GHz, this would correspond to a distance of merely 15 cm, which is indeed small compared to the distances between for example base stations and mobile terminals that are in the order of 100 m.

has been modelled by various statistical distributions. The most common of these are the Rayleigh and Rice models as follows:

▶ **Rayleigh distribution**
The channel h_{small} is defined in baseband notation in the form of a complex random variable as follows:

$$h_{small} = h_{re} + jh_{im}. \tag{4.8}$$

When the signal arrives at the receiver through many different paths of approximately the same power, the real and the imaginary components of the resulting sum field, h_{re} and jh_{im} respectively, are essentially sums of identically distributed random quantities. By the central limit theorem, the distribution of the sum of identically distributed random variables, follows a Gaussian distribution, and therefore the total received signal has Gaussian distributed real and imaginary components $h_{re}(t)$ and $h_{im}(t)$. Each one of them is a random variable with mean zero and variance $\sigma_m^2/2$, that is,

$$E[h_{re}] = 0, E[h_{re}^2] = \frac{\sigma_m^2}{2} \\ E[h_{im}] = 0, E[h_{im}^2] = \frac{\sigma_m^2}{2}. \tag{4.9}$$

The complex form of the channel coefficient can be found by

$$h_{small} = |h_{small}|e^{j\phi_{small}} \tag{4.10}$$

By transformation of variables, one can easily find that the magnitude of h_{small}, $|h_{small}| = \sqrt{h_{re}^2 + h_{im}^2}$ follows the Rayleigh distribution, and that the phase $\phi_{small} = \tan^{-1}(\frac{h_{im}}{h_{re}})$ follows the uniform distribution. The corresponding probability density functions are:

$$Pr\{|h_{small}|\} = \left(\frac{|h_{small}|}{\sigma_m^2}\right) e^{-\frac{|h_{small}|^2}{\sigma_m^2}} \tag{4.11}$$

$$Pr\{\arg(h_{small})\} = \frac{1}{2\pi}, -\pi \leq \arg(h_{small}) \leq \pi. \tag{4.12}$$

The average power of the channel is $E[|h_{small}|^2] = E[h_{re}^2] + E[h_{im}^2] = \sigma_m^2$. It is predicted from the path loss and the shadowing, that is, $\sigma_m^2 = h_{pl}h_{LARGE}$ (note that the shadowing is taken in linear form).

▶ **Ricean distribution**
Let us now assume that the signal is received through many scattered paths as before and one path from a single angle that is significantly stronger than the rest.

Commonly, this significant component will be the line of sight (LOS) component, that is, the one that travels along the direct path between the transmitter and the receiver. The real and imaginary components of the line of sight part now have nonzero mean real and imaginary values, $h_{LOS,re}$ and $h_{LOS,im}$ respectively, added to the zero mean Gaussian part calculated from the rest of the paths. Tthe equation for h_{small} now becomes

$$h_{small} = (h_{re} + h_{LOS,re}) + j(h_{im} + h_{LOS,im}).$$ (4.13)

The magnitude is therefore: $|h_{small}| = \sqrt{(h_{re} + h_{LOS,re})^2 + (h_{im} + h_{LOS,im})^2}$. The dominance of the significant path is usually measured by the Ricean K-factor K_f, which is is defined as the ratio of the power of the constant part of the signal over the average power of the random part of the signal.

$$K_f = \frac{|h_{LOS,re}|^2 + |h_{LOS,im}|^2}{\sigma_m^2}$$ (4.14)

σ_m^2 is the variance for a Ricean distribution. Thus the higher the Rice factor, the lower the power in the random part. A Rice factor of zero corresponds to Rayleigh distribution, while an infinite rice factor corresponds to no scattering and only a single LOS path. The resulting distribution for the envelope (magnitude), $|h_{small}|$ is the Ricean distribution which is given by:

$$Pr\{|h_{small}|\} = \left(\frac{|h_{small}|}{\sigma_m^2/2}\right) e^{-\frac{|h_{small}|^2}{\sigma_{rice}^2}} e^{-K_f} I_0 \left(\frac{|h_{small}|\sqrt{2K_f}}{\sigma_m/2}\right)$$ (4.15)

$I_0(\cdot)$ is the zero order Bessel function. In the case where the Ricean K-factor is zero ($K_f = 0$), then the equation reduces to the Rayleigh distribution.

Path loss modelling

In a channel, the large scale fading components need to be considered in terms of path loss and shadowing, while the small scale components need to be expressed as fading quantities as follows:

▶ Before considering any obstructions on the ground as well as the ground itself, the path loss between a transmitter and receiver depends on the distance spaced apart between the transmitter and receiver, r, the wavelength λ and the antenna gains at both ends, G_{Tx} and G_{Rx}, using the following derivation:

$$G_{Tx} G_{Rx} \left(\frac{\lambda}{4\pi r}\right)^2$$ (4.16)

If the ground is present, over suitably large distances, the path loss is then no longer dependent on wavelength but only on the heights of the transmitter and receiver, l_{Tx}^2

Path loss modelling (Continued)

and l_{Rx}^2 as well as the distance r:

$$G_{\mathrm{Tx}}G_{\mathrm{Rx}}\left(\frac{l_{\mathrm{Tx}}^2 l_{\mathrm{Rx}}^2}{r^4}\right) \tag{4.17}$$

However, with shadowing objects present, a more general path loss model approach needs to be applied where a reference distance, r_{break}, will have a given loss value and then it will decay over at a greater distance r due to a path loss exponent, γ:

$$G_{\mathrm{Tx}}G_{\mathrm{Rx}}\left(\frac{\lambda}{4\pi r_{\mathrm{break}}}\right)^2 \left(\frac{r_{\mathrm{break}}}{r}\right)^{\gamma} \tag{4.18}$$

▶ In the case of a nonline of sight, there will be small scale fading, h_{small} that follows the Rayleigh distribution, which will have a magnitude determined by the following probability density function and standard deviation, σ_m:

$$\left(\frac{|h_{\mathrm{small}}|}{\sigma_m^2}\right) e^{-\frac{|h_{\mathrm{small}}|^2}{\sigma_m^2}} \tag{4.19}$$

while the phase has a uniform distribution.

▶ In the case of a line of sight, there will in general be small scale fading that follows a Rice distribution and it will have the following probability density function dependent on Rice factor, K_f and zero order Bessel function $I_0()$:

$$\left(\frac{|h_{\mathrm{small}}|}{\sigma_m^2/2}\right) e^{-\frac{|h_{\mathrm{small}}|^2}{\sigma_{\mathrm{rice}}^2}} e^{-K_f} I_0\left(\frac{|h_{\mathrm{small}}|\sqrt{2}K_f}{\sigma_m/2}\right) \tag{4.20}$$

The Rice factor is a ratio of the power in the constant part to the power in the random part of the channel, when Rice factor is zero, there is a Rayleigh channel, while when it is infinite, there is perfect line of sight or free space conditions.

4.1.2 Models for the Prediction of the Temporal Variation of the Channel

Let us look at Figure 4.2. A mobile terminal is moving along the direction of the velocity vector $\overrightarrow{v}_m$. An incoming plane wave of carrier frequency $f_c = c/\lambda$ and unit amplitude is impinging on the terminal from an angle α relative to the direction of motion (c is the speed of light, λ is the wavelength). Let's assume that at time t, the mobile terminal is at the point A in the figure and the received signal is $\cos(2\pi f_c t + \phi_0)$, where ϕ_0 is some initial phase. At time Δt later, the mobile terminal has moved by $v_m \Delta t$ to the point B in the figure. The corresponding difference Δl in the direction of the wave propagation is

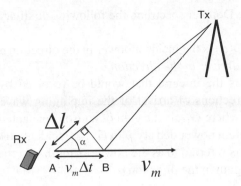

Figure 4.2   Explanation of the Doppler shift.

$\Delta l = v_m \Delta t \cos(\alpha)$. The received signal is now $\cos(2\pi f_c(t + \Delta t) + \phi_0 + \Delta\phi)$, where the change difference $\Delta\phi$ can be calculated by:

$$\Delta\phi = \frac{2\pi}{\lambda} v_m \Delta t \cos(\alpha). \tag{4.21}$$

So the mobile terminal perceives the incident wave as a wave of frequency:

$$f = \frac{\Delta\phi}{2\pi\Delta t} = \frac{1}{2\pi} \frac{2\pi f_c \Delta t + \frac{2\pi}{\lambda} v_m \Delta t \cos(\alpha)}{\Delta t} = f_c + f_c \frac{v_m}{c} \cos(\alpha). \tag{4.22}$$

Therefore the incident wave appears to have undergone a frequency shift by $f_c \frac{v_m}{c} \cos(\alpha)$. This phenomenon is called the Doppler phenomenon, and the frequency shift is called the Doppler shift $f_d = f_c \frac{v_m}{c} \cos(\alpha)$. The maximum Doppler shift is achieved when $\cos(\alpha) = 1$, or equivalently when $\alpha = 0$ and the receiver is moving towards the signal source, and is given by:

$$f_{d,\text{MAX}} = \frac{v_m}{c} \tag{4.23}$$

Assuming that the incident multipath components all have the same frequency f_c (i.e. they have resulted from reflection/scattering etc., from static objects and therefore have not undergone any additional Doppler shifts), the observed Doppler components are constrained to the frequency interval $f_c \pm f_{d,\text{MAX}}$.[6]

Clearly, as the different incoming paths have different incidence angles, they have different Doppler shifts. The effect of the shift of all the multipath components is described by the Doppler spectrum, $S(f)$, which describes the power density of the multipath components that have Doppler shifts around the frequency f.

---

[6] In reality, moving objects between the transmitter and receiver can cause components to appear outside of this range.

In order to derive the Doppler spectrum, the following auxiliary quantities are defined:

► The incident power $p(\alpha)$ is the incident power in the direction α, *that is, the fraction of the total power that arrives in the direction α.*
► The total power P is the integral that would be received by an isotropic antenna, summed over all directions of arrival of the impinging waves $\alpha \in [-\pi, \pi]$, *that is,* $P = G_{\text{iso}} \int p(\alpha)d\alpha$, where G_{iso} is the gain of an isotropic antenna,
► The normalised incident power density $p'(\alpha)$ is the incident power density in the direction α ($p'(\alpha) = \frac{p(\alpha)}{P}$ is referred to as the power azimuth spectrum).[7]
► $G(\alpha)$ is the antenna gain in the direction α.[8]

Let $f_d(\alpha)$ denote the Doppler shift that is caused by an incident wave from the direction α:

$$= f_c \frac{v}{c}\cos(\alpha), (df_d = -f_c \frac{v}{c}\sin(\alpha)d\alpha) \tag{4.24}$$

Clearly $f_d(\alpha) = f_d(-\alpha)$ and the waves incident from the angles in the ranges $\left[\alpha - \frac{d\alpha}{2}, \alpha + \frac{d\alpha}{2}\right]$ and $\left[-\alpha - \frac{d\alpha}{2}, -\alpha + \frac{d\alpha}{2}\right]$ have the same range of absolute values of Doppler shifts $|df_d|$.

Therefore the incident waves received from the angles in these two ranges contribute to the Doppler spectrum at the same value of the Doppler frequency f_d:

$$S(f_d)|df_d| = \left[p(\alpha)G(\alpha) + p(-\alpha)G(-\alpha)\right]|d\alpha|. \tag{4.25}$$

Additionally, it can be derived from (4.24) that:

$$|df_d| = \sqrt{f_{d,MAX}^2 - f_d^2}|d\alpha|. \tag{4.26}$$

Therefore the Doppler spectrum is:

$$S(f_d) = \frac{1}{\sqrt{f_{d,d,MAX}^2 - f^2}}\left[p(\alpha)G(\alpha) + p(-\alpha)G(-\alpha)\right] \tag{4.27}$$

where $\alpha = \cos^{-1}\left(\frac{f_d}{f_{d,d,MAX}}\right)$ and $S(f_d) = 0$ for $|f_d| > f_{d,d,MAX}$.

---

[7] This means that the power arriving from the direction α is $P\,p'(\alpha) = p(\alpha)$.
[8] This means that the power that the not necessarily isotropic receiving antenna receives from the direction α is $Pp'(\alpha)G(\alpha)$.

Specifically for the case where the gain is a constant for all angles $G(\alpha) = G$ and the same amount of power arrives from all directions $f(\alpha) = \frac{1}{2\pi}$ (omni-directional antenna and uniformly distributed angle of arrival for the incoming power), the Doppler spectrum from Equation (4.27) is the well-known bathtub spectrum and can be written as:

$$S(f_d) = \frac{PG}{\pi \sqrt{f_{d,d,MAX}^2 - f_d^2}} \qquad (4.28)$$

The Doppler spectrum bath tub model for a Rayleigh environment is shown in Figure 4.3. The curve is compared with a histogram of power of a simulated narrowband Rayleigh fading signal and it can be seen clearly that when assuming no scatterers are moving, there is negligible power both above maximum Doppler shift or below the minimum Doppler shift. In reality we would not see infinite power at the maximum or minimum Doppler shift so the bathtub curve is therefore valid within the Doppler bounds. Outside of these bounds we would in the case of measured data see a thermal noise floor.

One should note that in the case of a Ricean distribution, we would see a delta function spike at a frequency corresponding to the angle of the line of sight. The higher the Rice factor, the higher the magnitude of the delta function would be because it is directly proportional to the square root of the Rice factor (Saunders and Aragon-Zavala 2007). In different environments, different Doppler spreads are found. For example in

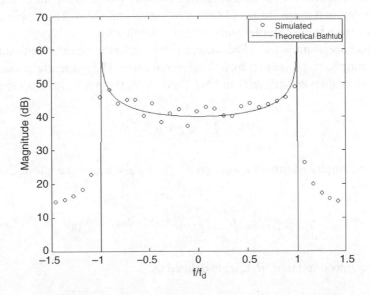

Figure 4.3 Diagram illustrating the bath tub Doppler spread model, compared with the Doppler spread of a simulated narrowband channel.

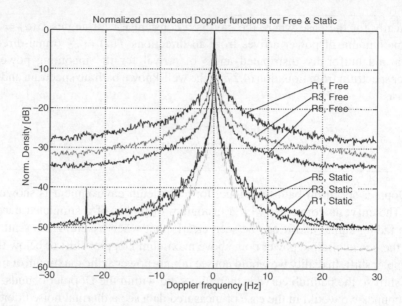

Figure 4.4 Diagram showing experimental results of the Doppler spread of a static laptop computer with the user interaction and the increase in Doppler spread. Source: Bach Andersen *et al.* (2009). Reproduced by permission of © IEEE 2009.

Figure 4.4, the Doppler spread of a measured channel where the mobiles are stationary laptop computers and there are people are moving around them. Thus there are moving scatterers causing a small Doppler shift, with a mean of zero.

The Doppler spectrum is useful because it is related to the autocorrelation function of the channel impulse response $h(t)$ by a Fourier transform. The autocorrelation function of $h(t)$ is defined by first deriving $h(t)$ and $h(t + \Delta t)$ from the power azimuth spectrum:

$$h(t) = \int_{-\pi}^{\pi} \sqrt{p(\alpha)} e^{j\psi(\alpha)} d\alpha \qquad (4.29)$$

$\psi(\alpha)$ is the random phase term for a wave arrival from a given arrival angle α. Consequently we also have:

$$h(t + \Delta t) = \int_{-\pi}^{\pi} \sqrt{p(\alpha)} e^{j(\phi(\alpha)) + \frac{2\pi}{\lambda} v_m \Delta t \cos\alpha} d\alpha. \qquad (4.30)$$

Therefore the autocorrelation, $R_h(\Delta t)$ is defined as:

$$R_h(\Delta t) = E[h(t)h^*(t + \Delta t)] =$$
$$E[\int_{-\pi}^{\pi} \int_{-\pi}^{\pi} \sqrt{p(\alpha)} \sqrt{p(\alpha')} e^{j\phi(\alpha)} e^{-j(\phi(\alpha')) + \frac{2\pi}{\lambda} v_m \Delta t \cos\alpha'} d\alpha d\alpha'] \qquad (4.31)$$

which simplifies to:

$$R_h(\Delta t) = \int_{-\pi}^{\pi} p(\alpha) e^{-j\frac{2\pi}{\lambda} v_m \Delta t \cos(\alpha)} d\alpha \qquad (4.32)$$

Since the Doppler shift is given by $f_d = \frac{v_m}{\lambda}\cos\alpha$, by introducing a change of variables and taking into account equation 4.27 (for simplicity we are assuming that the gain is the same for all directions) the autocorrelation can be written as:

$$R_h(\Delta t) = \int_{-\pi}^{\pi} p(\alpha) e^{-j2\pi f_d \Delta t} d\alpha = \int_{-f_{d,\max}}^{f_{d,\max}} S(f_d) e^{-j2\pi f_d \Delta t} df_d \qquad (4.33)$$

Thus autocorrelation, $R_h()$, depends on the degree of angular spread, such that wider angular spread will result in more rapid decrease in autocorrelation and the channel is more time variant. Alternatively, autocorrelation depends on the degree of Doppler spread, such that wider Doppler spread will result in more rapid decrease in autocorrelation and the channel is more time variant.

It should be noted that a time variant channel is equivalent to a space variant channel, if the mobile is moving at velocity v_m. The channel would change within the time Δt over the equivalent spatial displacement $\Delta r = v_m \Delta t$.

The autocorrelation determines the rate of change of the channel. If for small Δt, we have $R_h(\Delta t)$, then the channel has gone between two statistically independent realisations within this small time interval Δt. A channel *without memory* is a channel where the channel state at a particular point in time is independent of the channel at a previous state or next state, assuming that the states are spaced by delay Δt. In a real channel scenario, the change in environment will always mean that the channel state at a particular point has a relation to the previous or next channel state and therefore there will be a finite autocorrelation between the two channel states. In this scenario the channel will be *with memory*. The channel without memory would not have a defined Doppler spectrum and angle of arrival. In some applications it is enough to just model the channel without memory, which is known as a *Monte Carlo* simulation, where as in the case where we want to include the channel with memory, we must use a *correlated samples* simulation. We will now look at these two cases with appropriate MATLAB examples.

▶ Monte Carlo simulations

In some cases, it is sufficient to perform Monte Carlo simulations, that is, to generate several independent realisations of the channel with these known statistics. For example, if the goal were to generate $N = 1000$ independent realisations of a Rayleigh fading channel with unit average power, the following MATLAB command would suffice:

```
h = (randn(N, 1) + j*randn(N, 1))/ sqrt(2);
```

▶ Correlated samples

In several cases, it is necessary to simulate the temporal evolution of the channel so that it reflects not only the distribution but also the actual autocorrelation properties. Let us again assume that the goal is to generate $N = 1000$ independent realisations of a Rayleigh fading channel with unit average power so that they correspond to the wireless channel sampled every Δt and they have a known autocorrelation function $R_h(\Delta t) = E\left[h(u)h^*(u + \Delta t)\right]$. As a first step we generate N independent realisations of a Rayleigh fading channel with unit average power using the command as with a Monte Carlo simulation:

```
h = (randn(N, 1) +  j*randn(N, 1))/ sqrt(2);
```

Then we can pursue a time based or a frequency based filtering approach:

4.1.2.1   Time Based Filtering

Let us create the $N \times N$ autocorrelation matrix $\boldsymbol{R}$

$$
\boldsymbol{R} = \begin{bmatrix}
R(0) & R(\Delta t) & R(2\Delta t) & .. & R((N-1)\Delta t) \\
R(-\Delta t) & R(0) & R(\Delta t) & .. & R((N-2)\Delta t) \\
.. & .. & .. & .. & .. \\
R((1-N)\Delta t) & R((2-N)\Delta t) & R((3-N)\Delta t) & .. & R(0)
\end{bmatrix}
$$

$$(4.34)$$

Notice that the matrix $\boldsymbol{R}$ is a matrix that is complex conjugate symmetric, and has a Toeplitz structure. Therefore it can be easily constructed in MATLAB using the command (assuming the autocorrelation function $R(x)$ and Δt are already implemented):

```
t = (0:(N-1))*Delta_t;
row1=  R(t);
column1 = row1';
matrixR = toeplitz(row1, column1);
```

Let us now define the matrix $\boldsymbol{W} = \boldsymbol{R}^{1/2}$, that is, a matrix such that $\boldsymbol{W}\boldsymbol{W}^H = \boldsymbol{R}$. Let us now create the vector $\mathbf{w} = \boldsymbol{W}\mathbf{h}$.

```
W = sqrtm(matrixR);
w=W*h;
```

It can easily be shown that the elements of w are Gaussian circularly symmetric distributed (and therefore have Rayleigh distributed envelope), and have the desired correlation properties.

4.1.2.2 Frequency Based Filtering

We can also apply filtering in the Doppler frequency domain as follows. Let us first set a time step Δt that determines the maximum frequency (equal to the inverse of Δt) in a fast Fourier transform. Although the range of frequencies is determined by Δt, the number of time steps determines the fineness of the Fourier transform. Next we generate the Doppler spectrum **s** sampled at $n\Delta f$ (i.e. N points for N samples), where n ranges from $-N/2$ to $N/2$ and $\Delta f = 1/(N\Delta t)$. For simplicity we take the example of the bathtub spectrum:

```
freq = [div(N,2):1:div(N,2)]./(N*Delta_t);
s = P .* G ./ (pi .* sqrt((f_d_max^2) - (freq.^2)));
```

As the input channel, we generate N white complex Gaussian samples in the frequency domain in a vector **h** just as we would do for Monte Carlo though note that they are frequency samples and not time samples in this instance:

```
H = (randn(N, 1) + j*randn(N, 1))/ sqrt(2);
```

Finally we multiply **h** element wise with **s** and convert **h** to the time domain by using an inverse fast Fourier transform, hence the upper case notation H:

```
H = H.*s;
h = ifft(H);
```

Channels with and without memory

The small scale components of a channel can be modelled in the simplest way by a random noise signal, which has a defined statistical distribution such as a Rice distribution or Rayleigh distribution. The instantaneous value of this small scale fading (which can be considered as the channel state) will have some dependency on the channel state in previous time instances. Furthermore the next channel state will have some dependency on the current channel state and those before it. Therefore a channel that maintains this dependency is known as a channel with memory. If a channel does not have memory, then the current channel state is completely random and has no relation to the previous channel states.

In order for the small scale fading in a channel to have memory, it will require a defined Doppler spread, thus meaning that a channel with memory will be composed as follows:

► The noise signal representing the small scale fading will consist of a set of components that have their individual Doppler frequency, f_d, which is determined by

Channels with and without memory (Continued)

$(v_m/c)f_c\cos\alpha$, where v_m is the mobile velocity, c is the speed of light and α is the angle at which that particular component is arriving at the mobile at frequency f_c relative to its direction.

▶ The combination of the separate components at their respective Doppler frequencies will form a Doppler spectrum.

▶ The temporal (or spatial) variation of the channel is described by the Doppler spectrum, which therefore depends on the angular spread of the multipath components.

▶ The Doppler spectrum is the Fourier transform of the channel autocorrelation that captures the similarity of channel realisations at neighbouring points in time (or space).

Therefore another means by which a channel without memory can be described is one with independent identically distributed realisations. A channel with memory is one with correlated realisations. Regardless of whether the channel has memory or not, it still holds a Rayleigh or Ricean distribution.

4.1.3   Narrowband and Wideband Channels

So far we have considered the input $x(t)$ and the output $y(t)$ to be at a single frequency, which means that the channel, $h(t) = h.\delta(t)$, is a scalar. If the bandwidth of $x(t)$ were larger but the channel coefficient were constant over it, then the same description would apply. However, in the general case, we will need to caption the dependence on frequency, that is, characterise the channel over a wide band. This section will therefore look in more detail into how to determine whether a channel is considered narrowband or wideband, as well as how a wideband channel is analysed.

First of all, we will need to express $h(t)$ in more detail than we have done so far. In the general case, multiple delayed copies of the signal are received at different delays τ and the channel itself is different at different observation times t as the user is moving. The simplest case of this is seen in the top of Figure 4.5, where there is a direct path at a given instant, which arrives at delay τ_0 and there is a single scatterer that causes another path to arrive later at delay τ_1. There could be up to N scatterers, where of course there would result in being τ_1 up to τ_N delays though τ_0 would only exist if there was a line of sight. Therefore, the effect of the channel can be considered as that of a linear filter response $h(t, \tau)$ of which varies according to the observation time t (i.e. the channel is a time variant filter) and on the delay τ. The received signal $y(t)$ is determined as a convolution of $x(t)$ and $h(t, \tau)$ as follows:

$$y(t) = h(t, \tau) * x(t) + n(t) = \int_{-\infty}^{\infty} x(t - \tau)h(t, \tau)d\tau + n(t) \qquad (4.35)$$

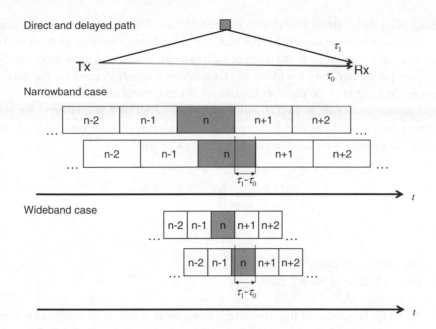

Figure 4.5   Illustration of creating a wideband channel by increasing the signal bandwidth, or reducing the symbol time.

where ∗ denotes the convolution operation and $n(t)$ is a noise component as used earlier in the chapter. The channel impulse response $h(t, \tau)$ is expressed in the baseband and is therefore a complex function.

Multipath components arriving at different delay times τ correspond to various routes between the transmitter and the receiver. These different routes have different lengths and therefore the signal components are expected to arrive at the receiver with different delays. Moreover they have undergone different attenuations depending on the materials that the wave interacted with between the transmitter and the receiver. What the receiver observes is the superposition of these delayed and scaled components (i.e. the superposition of delayed and scaled versions of the transmitted data stream, which is composed of a sequence of symbols).

Up to this point, it has been assumed that all the multipath components arrive with delays τ_1 to τ_N that are very close to each other relative to the duration of a single symbol and therefore we only need to consider the addition of the signal components. Figure 4.5 illustrates this for the case of just one scatterer where the difference $\tau_1 - \tau_0$ is small compared to the symbol length represented as blocks in the first instance. When these conditions have been met, then it is considered a narrowband channel.

As the symbol time of the transmitted signal decreases (or equivalently the signal bandwidth increases), the effect of the multipath component causes symbol n from the delayed path to almost overlap symbol $(n + 1)$ of the direct path. In this instance, the delay is significant enough compared to a symbol length that the channel is considered wideband. The same effect could also happen if the symbol length was kept constant (i.e. the bandwidth was constant) but the scatterer was placed such that its delay increased far enough to cause symbol n of the delayed path to again almost overlap with symbol $n + 1$ of the direct path. Therefore a wideband system does not only depend on the bandwidth of the radio link but also the channel conditions for that particular radio link.

Let us now consider these two situations in more detail using equations. We start by considering the channel impulse response for a fixed value of t, so we now only need to define $h(\tau)$ as follows which is a summation of delay paths:

$$h(\tau) = \sum_{n=0}^{N_t-1} a_n e^{j\phi_n} \delta(\tau - \tau_n), \tag{4.36}$$

where

▶ N_t is the number of channel paths,
▶ a_n is the amplitude of the nth path,
▶ ϕ_n is its phase, and
▶ τ_n is its delay. The delay of the first path is commonly taken to be equal to zero and the delay of each of the following paths is frequently referred to as excess delay.

We distinguish two cases:

1. Narrowband channels
   When the symbol time is very large relative to the temporal extent of the channel, then the effect of the delayed taps is confined within the duration of a single transmitted symbol and the equivalent channel impulse response is a single delta function:

$$h_{\text{NB}}(\tau) \approx \left(\sum_{n=0}^{N_t-1} a_n e^{j\phi_n} \right) \delta(\tau) \tag{4.37}$$

   In the frequency domain, the Fourier transform of such a function would be constant over all frequencies. Such a channel is referred to as a *narrowband* or a *frequency flat* channel.
2. Wideband channels
   When the symbol time is smaller than the temporal extent of the channel, then the effect of the delayed components extends over several transmitted symbols.[9]

$$h_{\text{WB}}(\tau) = \sum_{n=0}^{N_t-1} a_n e^{j\phi_n} \delta(\tau - \tau_n), \tag{4.38}$$

---

[9] The effect of previously transmitted symbols that are superimposed on the current symbol because of delayed copies received is known as *intersymbol interference* (ISI).

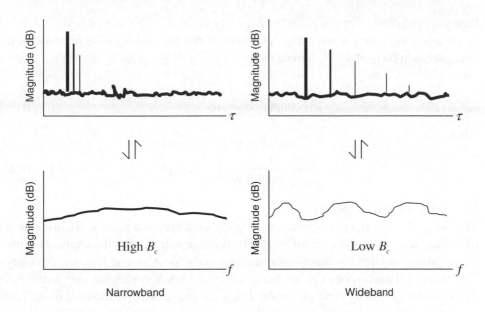

Figure 4.6   Comparison of the narrowband and wideband fading channel.

In the frequency domain, the Fourier transform of such a function would be variable over frequency. Such a channel is referred to as a *wideband* or a *frequency selective* channel.

The difference between narrowband and wideband channels is roughly illustrated in Figure 4.6, where if an impulse is transmitted through a radio channel, several copies are received with different attenuations. The later impulses arrive due to the different longer paths along which the signal has travelled. When the narrowband and wideband signals are looked at in the frequency domain (by applying a Fourier transform) the channel does not change significantly with frequency, where as in the wideband case, there is a great change, thus there is frequency selectivity.

At this point it is useful to introduce the notion of the *power delay profile* (pdp(τ)). The power delay profile is calculated as the expected value of the power of the channel impulse response over the local area statistics and is a function of the delay τ.

$$\text{pdp}(\tau) = E_t \left[|h(t, \tau)|^2 \right] \qquad (4.39)$$

It can be considered as a discrete function of the delay τ (when the channel impulse response is given as a sum of delayed components as in Equation (4.38)) or as a continuous function of the delay τ. The frequency selectivity of the channel is characterised by different metrics. In this section, we discuss two commonly used metrics, namely the *delay spread* and the *coherence bandwidth*.

▶ Delay spread (DS)
A common measure for the temporal extent of the channel impulse response is the
delay spread (DS), which is defined as:

$$DS^2 = \frac{1}{\int_{-\infty}^{+\infty} pdp(\tau)d\tau} \int_{-\infty}^{+\infty} (\tau - \bar{\tau})^2 \, pdp(\tau)d\tau \qquad (4.40)$$

where

$$\bar{\tau} = \frac{1}{\int_{-\infty}^{+\infty} pdp(\tau)d\tau} \int_{-\infty}^{+\infty} \tau pdp(\tau)d\tau \qquad (4.41)$$

One can think of it as follows:
The integral $\int_{-\infty}^{+\infty} pdp(\tau)d\tau$ corresponds to the total received energy. By normalising
(dividing) the power delay profile $pdp(\tau)$ by this integral, we are effectively converting
into a probability density function (a function that is positive and integrates to unity).
The quantity $\bar{\tau}$ corresponds to the mean value of a random variable that would have
this probability density function, while DS is the second central moment of the same
random variable.

▶ b. Coherence bandwidth B_c
Qualitatively speaking, the coherence bandwidth B_c of the channel is the bandwidth
over which the channel transfer function is approximately constant. Quantitatively, in
order to find the coherence bandwidth we need to look at the frequency coherence
function and define a certain level.

Let us consider the two-dimensional expression of the channel impulse response
$h(t, \tau)$, and its corresponding Fourier transform with respect to τ, $H(t, f)$. The wireless
channel is commonly assumed to be a wide sense stationary process. Also if the channel taps are zero mean random variables (like they would be in the case of Rayleigh
distributed taps), then $E_t[H(t, f)] = 0$, where again $E_t[\cdot]$ indicates averaging over
time, that is, over the local statistics. The frequency autocorrelation function $R(\Delta f)$ is
defined as:

$$R(\Delta f) = E_t[H(t, f)H^*(t, f + \Delta f)]. \qquad (4.42)$$

The frequency autocovariance function $R_f(\Delta f)$ is the same as the frequency autocorrelation function because $E_t[H(t, f)] = 0$, with zero mean,[10] and takes its maximum
at $\Delta f = 0$:

$$R_f(\Delta f) = R_f(\Delta f), |R_f(\Delta f)| \leq R(0) \qquad (4.43)$$

Clearly both the autocorrelation and the autocovariance functions are complex conjugate symmetric. The normalised autocovariance function is derived by dividing

---

[10] This is why people frequently refer to the autocorrelation, when they really mean the autocovariance

(normalising) the autocovariance function by its maximum value and therefore it is given by:

$$R_{f\text{norm}}(\Delta f) = \frac{R_f(\Delta f)}{R_f(0)} \tag{4.44}$$

The coherence bandwidth at level R_{coh}, defined as $B_{\text{coh}}(R_{\text{coh}})$ is calculated as the bandwidth beyond which the normalised frequency autocovariance function falls below the level R_{coh}. Common values for R_{coh} are 0.99, 0.95 and 0.9, which are high values that indicate a coherence bandwidth within which the channel stays highly correlated.

The delay spread and the coherence bandwidth are related to each other. Intuitively, the larger the delay spread, the longer the significant temporal extent of the channel and the more frequency selective the channel will be (faster variation in frequency), which would mean that the coherence bandwidth is low. The mathematical connection between the two metrics arises from the fact that the autocorrelation function is the Fourier transform of the power delay profile. Although no exact relationship exists between the delay spread and the coherence bandwidth a good rule of thumb is to take $B_c \geq \frac{1}{10\text{DS}}$ where the chosen correlation threshold is 0.8 (Fleury 1996).

Another good rule of thumb relates the coherence bandwidth of the channel B_c to the system bandwidth B in order to determine whether the channel is frequency selective or not. If $B \gg B_c$ or equivalently if $T_s \ll \text{DS}$ (which in practice means if $B > 10B_c$ or equivalently $T_s < \frac{\text{DS}}{10}$), then the channel is considered frequency selective (i.e. wideband). If $B \ll B_c$ or equivalently if $T_s \gg \text{DS}$ (which in practice means if $B < \frac{B_c}{10}$ or equivalently if $T_s > 10\text{DS}$), then the channel is considered frequency flat (i.e. narrowband). If a channel has a coherence bandwidth (or equivalently delay spread) between these two boundaries then it is on the borderline between narrowband and wideband.

Narrowband and wideband channels

Channels are described as either *wideband* or *narrowband*, which will be dependent both on the system bandwidth and the radio environment that consists of paths with different multipath delays through which the same set of data symbols (each with the same symbol time) are transmitted and received. The following steps will explain their differences:

▶ A narrowband system is one where all the multipath components arrive within delays that are negligible with respect to the symbol time. In order for the symbol time to be long enough in this instance, then the bandwidth will be low. Therefore the system is narrowband.

Narrowband and wideband channels (Continued)

▶ A wideband system is one where the multipath components arrive with delays that are comparable to the symbol time. In order for the symbol time to be short enough in this instance, then the bandwidth will be high. Therefore the system is wideband.

Any narrowband or wideband system will have a power delay profile, which defines how much power is received at each delay, forming a *delay tap*. Each individual delay tap can be Rayleigh or Ricean distributed. The change in time between the delay taps and the magnitude of the maximum time delay will enable the *delay spread* to be derived, which is the root mean square of the delays present. This is a useful quantity to measure the temporal extent of the channel. Therefore there are two more points to note about the difference between narrowband and wideband channels:

▶ A small delay spread will mean the system is narrowband because it is receiving small delays compared to the symbol rate. Therefore in the frequency domain the channel will change little, thus **a narrowband system has a high coherence bandwidth**.
▶ A high delay spread in the system will mean that the system is wideband because the system will receive large multipath delays within the symbol time. Therefore in the frequency domain the channel will change rapidly, thus **a wideband system has a low coherence bandwidth**.

The coherence bandwidth defines the stability of the channel in the frequency domain. The larger the delay spread, the smaller the coherence bandwidth. If the coherence bandwidth is small, the channel is therefore frequency selective.

4.1.4 Polarisation

The electromagnetic waves transmitted by the transmitting antennas have certain polarisation characteristics depending on the type of antenna used and its orientation. Commonly, antennas are vertically polarised: think of a vertically oriented dipole as the transmitting antenna. As the signal gets reflected, diffracted and scattered, part of its energy stays in the original (e.g. vertical) polarisation, and part of it gets coupled into the other (e.g. horizontal). The receive antenna also has certain polarisation characteristics and can pick up one or the other or both polarisations (to different degrees).

For the purposes of considering polarisation in this book, the cross polarisation patio, XPR, is used, which is the ratio of the time averaged power received in the vertical polarisation P_V to the time averaged power received in the horizontal polarisation P_H.

Therefore:

$$\text{XPR} = \frac{P_V}{P_H} \tag{4.45}$$

4.1.5   Summary of Parameters Required for SISO Channel Modelling

For the SISO case, channel modelling has focused on the following attributes of the channel:

▶ **Power distribution**: The path loss models predict how the received power decays with distance between the transmitter and the receiver. The shadowing models describe the statistical distribution around the path loss for locations that are at the same distance from the transmitter for example Rayleigh or Rice fading distributions characterise the statistical properties of the narrowband fading on the channel over local areas.
▶ **The power delay profile**: It is applicable for wideband channels and describes how much the signal arrives at different delays. The power delay profile models have been made consistent with the experimental results derived through extensive channel measurements in different environments and at different frequencies.
▶ **Angular properties**: The angular properties of the channel determine the Doppler spectrum of the signal for a moving receiver/ transmitter and therefore are important for the description of time varying signals

4.2   Challenges in MIMO Channel Modelling

In the MIMO case, channel modelling should describe the links between all transmitting and receiving branches at one instant. Each link is in itself a SISO link.

In most MIMO systems, the multiple antennas at the receive (or transmit) end are assumed to be arranged in an array of some geometric configuration. Therefore all of them are assumed to be in the same local area, and therefore experience the same path loss and shadowing. The large scale dependencies which describe the average received power follow the same power loss and fading laws as for SISO communications, and hence they also apply in MIMO. Moreover, the signals on all the array antennas are assumed to have the same statistics, i.e., the same distribution, the same mean and the same variance. However, since they are at different spatial locations, they are assumed to experience different realisations of the small-scale fading, yet not independent of each other. The challenge in MIMO lies in the characterisation of the detailed co-dependence of the links.

A simple approach to MIMO modelling was to assume that all the links are independent identically distributed (i.i.d.) Rayleigh fading SISO channels. However this simple approach does not reflect reality and more evolved channel models had to be developed and validated through measurements and simulations. There is no one particular model that is the 'best' model. Different models have been developed for different purposes and vary in their complexity and the level of detail. Let us consider an example of a system where the goal is to evaluate a demodulation algorithm. In this case the channel model will need to reflect the details of the physical (PHY) layer so that it is then possible to predict for example the bit error rate of coding schemes that result from interdependence between symbols from different antennas. We may on the other hand be interested in a throughput metric, where the channel model would need to consider details of the physical layer that cause complexities in the medium access control (MAC) layer. The relevant metric in this instance would be the packet error rate that summarises the rate of successful transmissions over several realisations of the channel. Moreover, two channel models might have been created for the same purpose but they trade off computation versus accuracy.

Channel models are classified as *deterministic* or *stochastic*, depending on how they are constructed:

Deterministic Models

This class of channel models stems from the efforts to develop models that precisely reflect the propagation in the actual environment where the system is to be deployed. Within deterministic channel, we can further distinguish:

► Recorded channel responses
► Ray tracing based approaches

The derivation of deterministic models is time consuming and complex and they are often only used to cover small indoor environments.

Stochastic Models Stochastic models are based on purely statistical (random) generation of a channel for which certain parameters are known. A simple example is the i.i.d. Rayleigh channel whereby each of the paths within the MIMO link are generated as independent Rayleigh distributed random variables with such input parameters as the average power of the links (as the previous analysis showed, the Rayleigh distribution is a single parameter distribution and knowing the mean is enough to determine the variance etc.). However, stochastic models incorporate other factors to determine the small scale interdependencies between the MIMO paths. Within stochastic based models we can distinguish:

> *1. Geometrically based models:* This approach tries to create a channel model based on the geometric layout of the physical surroundings and as such can allow the model to better reflect the effect of the environment.
>
> *2. Parametric models:* This modelling method abstracts from the detailed interactions with objects in the environment, and approximates the complex interactions as waves with statistically known direction of arrival/departure (angular properties). It then adds them together to create a multipath channel. This approach often increases the complexity for suitable accuracy though its application can be powerful.
>
> *3. Correlation based models:* The additional input parameter that is used for generating the random realisations of the channel is a known value of the channel correlations, to introduce an interdependence. This is a typical example of a stochastic model where the known parameters include the distributions of the individual channel gains, their mean and variance, as well as their correlation.

4. Eigenmode driven models: By considering the eigenmodes of the channel and how they behave (rate of variation, co-dependence), the channel behaviour can be assembled in terms of its constituent equivalent spatial subchannels.

The following sections will give some examples of well known channel models and will move progressively from the very intuitive deterministic channel models to the more abstract stochastic based models.

4.2.1 Deterministic Models

We start with deterministic models because they are the easiest to understand. Two of the main kinds of deterministic based models are summarised here.

1. Recorded impulse responses

This type of model uses **channel measurement recordings**. Extensive measurements of MIMO channels are undertaken in different environments, from which actual data is recorded. The data reflect the recorded impulse responses for every measurement location and for each link, that is, for each combination of transmit-receive antennas. This method is not used frequently because the data collected are not necessarily representative of several locations or several environments, despite the fact that they are extremely accurate for the locations where the measurements were taken. Moreover this approach is very data-intensive, in the sense that large amounts of information need to be stored and processed.

2. Ray tracing

The other main form of deterministic modelling is that of **ray tracing**, which is based on geometrical ray optics (Ling *et al.* 2001).
    Ray tracing tools require that the following be described:

1. The environment, that is, the geometry and the composition of all the boundaries (e.g. walls) and objects (e.g. furniture).
2. The electromagnetic properties of the materials used for the boundaries and objects at the frequency of operation
3. The location and orientation of the Tx and the Rx antennas in this environment.
4. The antenna gain pattern for each Tx and Rx antenna.

For each link, all the paths that the signal can follow from the Tx to the Rx are calculated. Each one of them is considered a ray. The properties used are the following:

▶ Line of sight propagation,
▶ Rays reflected off objects in the environment one or several times. According to Snell's law, the incidence and reflection angles are equal and each reflected ray is attenuated

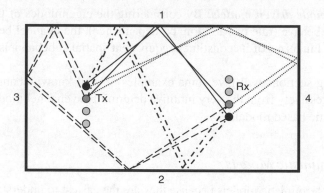

Figure 4.7   Simple example of a ray tracing model.

by the corresponding reflection coefficient which depends on the incidence angle and
the properties of the material that the ray is reflected from. Paths that correspond
to more than a prescribed maximum number of reflections are assumed to not have
significant contributions and are ignored. The surviving paths have different magnitudes
and different delays because of the different path lengths.

▶ Rays scattered off objects in the environment. In contrast to reflection scattering is what
is commonly happens when a ray hits an object with a rough surface and therefore the
rays that bounce off do not follow directions specified by Snell's law.

▶ Rays diffracted around objects in the environment.

The contribution from all the paths constitutes the wideband impulse response of the
channel between the Tx and the Rx. The method has its roots in SISO channels. In order to
produce the MIMO channel, the operation is repeated for all links, that is, for the calculation
of the paths between each transmit antenna and each receive antenna. A simple example of
how ray tracing operates is shown in Figure 4.7. The line of sight is not shown for clarity.
If the Tx and Rx positions with the darkest spot are considered first, there are four different
ray paths illustrated using different style lines. If we then focus on the neighbouring spots
at both the Tx and Rx, we can see that the corresponding four rays and how they have
slightly changed are also illustrated. Thus due to the proximity of the two spots at both
ends, the rays are changing only slightly in characteristics. Because of the minor change,
it creates an interdependence between all the possible Tx to Rx branches.

Ray tracing can provide highly accurate MIMO models and precisely capture the interdependencies between all the links. An example of such a tool is *WiSE* developed by
Alcatel-Lucent.[11] However, ray tracing is one of the most complex and data intensive methods of modelling requiring specific information for a given location about the geometry,

---

[11] www.alcatel-lucent.com

materials and the system dynamic range. Therefore such techniques are limited to indoor radio environments.

Deterministic models
Deterministic models are environment specific instances of a channel, calculated using physical input parameters or they measured directly. Such models will accurately determine the interdependence between different MIMO branches by considering the relationship that one MIMO branch has to the other due to them being spatially separated for example. For the environment in question such models have very precise accuracy and enable point to point comparisons to be made as a mobile moves from one position to another. However, at the same time they do not provide a model that is representative of other environments, which does not result in a generalised model to test the general performance of a MIMO system, only its performance in a small number of environments.

4.2.2   Stochastic Models

4.2.2.1   Geometrically Based Models

As we saw earlier, ray-tracing calculates all the paths that a signal can follow from the transmitter to the receiver as it is reflected from objects in the environment. Geometric stochastic models try to build an abstraction of the objects in the environment by considering them as scatterers with a certain geometric layout around the transmitter, around the receiver or both. The difference between a scatterer and a reflecting object is that a reflection uniquely and precisely specifies the direction into which the wave goes after interacting with the reflecting object, whereas scattering does not. Instead, the waves are assumed to scatter in all directions, possibly with a different gain in each direction. The random (stochastic) aspect is introduced by assuming that the location of the scatterers is not strictly specified as in the case of deterministic models. Instead it is assumed to be random, but with known statistical properties.

In order to understand how geometrically built models are constructed, geometrically based stochastic SISO models will first be explained. Let us first concentrate on the case of a SISO channel where a base station (Tx) equipped with a single antenna is communicating with a mobile station (Rx), also equipped with a single antenna. There are several scattering objects around the Rx as shown in Figure 4.8.

For clarity, we start by considering just one of the scatterers: the signal from the Tx travels along the line of length r_1 between the transmit antenna and the scattering object, hits the object and the signal gets scattered. The part of it that gets scattered in the direction of the receiver travels a distance r_2 and reaches the Rx. The contribution to the channel

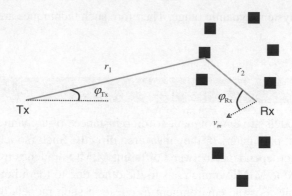

Figure 4.8 Diagram of the geometric arrangements of single scatterers.

impulse response from this scattering event is:

$$h(\tau) = L_{SCAT}\delta(\tau - \tau_{SCAT1}) \tag{4.46}$$

where τ_{SCAT} is equal to $(r_1 + r_2)/c$ and is the propagation time along the two paths, and L_{SCAT} is found by calculating the path attenuation in the two-step process. Assuming that the paths r_1 and r_2 are in free space, their path loss can be derived by a modification of the free space path loss equation:

$$L_{SCAT} = \sqrt{G_{Tx}(\phi_{Tx})G_{Rx}(\phi_{Rx})}k_r \left(\frac{\lambda}{4\pi}\right)\left(\frac{1}{r_1 + r_2}\right)e^{-j\beta(r_1+r_2)} \tag{4.47}$$

$G_{Tx}(\phi_{Tx})$, $G_{Rx}(\phi_{Rx})$ are the antenna gains of the Tx and the Rx in the directions ϕ_{Tx} and ϕ_{Rx} respectively, the angles ϕ_{Tx} and ϕ_{Rx} are calculated from the geometry, k_r is the scattering coefficient (it depends on the scatterer material, its geometric properties and the scattering cross-section), $\beta = \frac{2\pi}{\lambda}$ where λ is the wavelength. Note that L_{SCAT} incorporates the phase change.

Now let us include the other scatterers surrounding the mobile. Each one has a contribution of the previous form, and the complete channel is made up of the contributions from all the scatterers.

Given the different locations of the scatterers, the distances r_1 and r_2 will be different resulting in different loss coefficients L_{SCAT_l} and delays τ_{SCAT_l} for the lth scatterer. If N_s is the total number of scatterers, the complete SISO impulse response can then be found by:

$$h(\tau) = \sum_{l=1}^{N_s} L_{SCAT_l}\delta(\tau - \tau_{SCAT_l}) \tag{4.48}$$

As either end of the communication link moves, the contributions from the scatterers change in amplitude and phase and therefore the model becomes time variant. Let us assume that the Rx is moving. The impulse response at any given time instant, t, $h(t, \tau)$, or equivalent at the spatial location where the Rx is found at time t, can be written as:

$$h(t, \tau) = \sum_{l=1}^{N_s} L_{\text{SCAT}_l}(t)\delta(\tau - \tau_{\text{SCAT}_l}) \qquad (4.49)$$

where

$$L_{\text{SCAT1}rl}(t) = \sqrt{G_{\text{Tx}}(\phi_{\text{Tx}})G_{\text{Rx}}(\phi_{\text{Rx}})}k_r \left(\frac{\lambda}{4\pi}\right) \left(\frac{1}{r_{1l} + r_{2l}}\right) e^{-j\beta(r_{1l}+r_{2l})} e^{-j\beta v_m t \cos\varphi_{\text{inc},l}}. \qquad (4.50)$$

v_m is the velocity of the Rx motion.

The added phasor term can also be explained in terms of the plane wave approximation. When a plane wave is incident with an angle ϕ_{inc} on the line defined by two points A and B, the two points experience a phase difference that is equal to $\frac{2\pi}{\lambda}d_{\text{AB}}\cos\phi_{\text{inc}}$, where d_{AB} is the distance between the two points.[12]

The assumption underlaying Equation (4.50) is that the scatterers are some distance away from the Rx compared to the distance it has travelled in a given time window, so that the angle ϕ_{Rx} (angle of incidence of the ray relative to the direction of motion of the Rx) can be considered approximately constant.[13] This approximation is generally a good one when the distances to and from the scatterer, r_1 and r_2 respectively, are much larger than the distance travelled by the receiver antenna ($v_m t$).

Starting from the expression above for a SISO spatially-dependent channel impulse response, one can easily extend from the SISO to the MIMO case, by explicitly calculating the channel from each transmitter antenna to each receiver antenna as above. However, the calculation can be further simplified by assuming that the scatterers are sufficiently far from the Tx or Rx antennas so that the scattered waves appear like plane waves. Then the plane wave approximation can be used and the phase difference among the array elements at each end of the communication link can easily be calculated on the basis of the incidence angle relative to the element orientation.

For example let us consider the case of a system with two antennas at the Tx and two antennas at the Rx, separated by $d_{s,\text{Tx}}$ and $d_{s,\text{Rx}}$ respectively. For simplicity, we will also assume that the arrays are oriented along the y-axis although any orientation is easy to calculate as long as the incidence angle is correctly calculated. The geometry is shown in Figure 4.9.

Let $h_{ij}(t, \tau)$ denote the channel impulse response between transmit antenna j and receive antenna i. If Equation (4.49) is used to generate the channel gain $h_{11}(t, \tau)$ between transmit antenna 1 and receive antenna 1, using the plane wave approximation, the other channel

---

[12] The approach was explained in the context of the Doppler phenomenon.
[13] If the receiver moves a significant distance, the distances, angles and reflection coefficients will all change so the model parameters would therefore have to be updated.

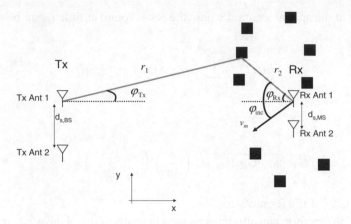

Figure 4.9 2×2 system in a single scattering situation.

coefficients can be found as:

$$h_{21}(t, \tau) = h_{11}(t, \tau)e^{-j\beta d_{s,Rx}\sin\varphi_{Rx}} \tag{4.51}$$

$$h_{12}(t, \tau) = h_{11}(t, \tau)e^{-j\beta d_{s,Tx}\sin\varphi_{Tx}} \tag{4.52}$$

$$h_{22}(t, \tau) = h_{11}(t, \tau)e^{-j\beta(d_{s,Rx}\sin\varphi_{Rx}+d_{s,Tx}\sin\varphi_{Tx})} \tag{4.53}$$

By using these equations it can be seen how there is an interdependence between the transmit-receive antenna combinations. This model assumes that the spacing between the antennas is negligible compared to the distance between the array and the scatterers. In the cases where the scatterers may be near, the model would have to be constructed with more complexity by considering each link independently. The stochastic nature of such channels is achieved by assuming that each realisation of the channel corresponds to a random placement of the scattering objects. Further realisations are generated by placing the scattering objects at different random locations which are chosen from known distributions (e.g. uniform over a disk, over a ring, over an ellipse etc.). In order to construct a geometric stochastic model, we need to parameterise by the following parameters:

▶ distance between the Tx and the Rx;[14]
▶ spacing between the antennas;
▶ terminal mobility;
▶ distribution of scatterers around the Tx and the Rx (distance, angle, density etc.);
▶ reflection coefficients (typical angles of incidence on each scatterer, typical materials);

---

[14] Only azimuth angles around the Tx and Rx have been considered. It is also possible that the Tx and Rx have different heights, which means that an elevation angle needs to be considered as well.

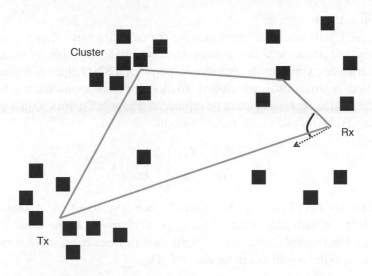

Figure 4.10   Illustration an example case of clusters.

▶ the center frequency of operation, which affects the wavelength λ and the wave-number β.

The reliability of such a channel model depends on the accuracy of the parameters involved. Extensive measurements and ray tracing simulations can be used to determine the values of the parameters. However, with appropriate parameter setting, geometrically built stochastic models can be applied to a wide variety of different environments.

There are further points we would need to add to the model to consider some necessary refinements being that of clustering, polarisation and diffuse scattering as follows:

1. Clustering of scattering objects
   A major aspect of geometric based modelling is the characterisation of channel clusters as illustrated in Figure 4.10. Clusters are considered as groups of scattering objects that are at a distance away from the Tx and Rx but still contribute a significant part to the channel impulse response. The underlaying assumption behind the inclusion of clusters is that scattering objects tend not to be uniformly distributed in space. This to a large extent reflects the physical picture observed when we look at our surroundings. Objects appear concentrated in some distinct locations (e.g. building blocks), and these areas are separated by empty space between them (e.g. parks, roads). In order to realistically include scattering clusters in simulated channels, the model input data would there-fore consider the location of such clusters and their characteristics. Additionally, the perceived clusters are time variant either because the mobile moves or the scattering objects themselves move. To reflect this phenomenon, cluster based models such as the COST 259 also specify the rate of appearance/disappearance of clusters as well (Correia *et al.* 2001).

2. Polarisation effects

Geometrically built stochastic models have the advantage that they can account for several physical phenomena in a mathematically simple way, like for example polarisation. Let us first consider the impulse response of a SISO channel represented in a way that both polarisations are included. To include polarisations, the link between the transmitter and the receiver, L, can be represented as a 2x2 matrix with 4 polarisation states using Equation (4.54) assuming isotropic antennas.

$$L = \begin{bmatrix} T_{VV} & T_{VH} \\ T_{HV} & T_{HH} \end{bmatrix} \left(\frac{\lambda}{4\pi r} \right) e^{-j\beta r} \tag{4.54}$$

The coefficients T_{VV}, T_{VH}, T_{HV}, T_{HH} describe how signals of each polarisation stay in the original polarisation state (T_{VV}, T_{HH}) or couple into the other polarisation (T_{HV}, T_{VH}). For example, in a line of sight case in free space, there is no polarisation coupling so the equation can be simplified to:

$$L_{LOS} = \begin{bmatrix} 1 & 0 \\ 0 & 1 \end{bmatrix} \left(\frac{\lambda}{4\pi r} \right) e^{-j\beta r} \tag{4.55}$$

In reality, the antennas used at each end will transmit/ receive on a combination of vertical and horizontal polarisations. The tap response $\mathbf{h}(\tau)$ in this case would be a delta pulse at delay τ_{LOS}, which is determined by the length of the line of sight path, and it would make up the entire channel impulse response.

$$\mathbf{h}(\tau) = L_{LOS}\delta(\tau - \tau_{LOS}) \tag{4.56}$$

($\tau_{LOS} = r/c$ and c is the speed of light).

3. Diffuse scattering

So far, the assumption has been that the scatterers cause only specular reflections. In some cases there can be instances of diffuse reflections, which are due to irregularities and roughness of the scattering objects as illustrated in Figure 4.11. If the surface of the reflecting object is flat and smooth (or has negligible roughness) then there will be just one reflection which is known as a single specular reflection. However, if there is a large degree of roughness, there will be lots of reflections that come from one single source and all arrive at the same time.

Based on a principle known as a Lambertian case (Degli-Esposti 2001), we can account for both the specular reflections and the diffuse scattering by defining the total scattering $L_{SCAT1dl}(t)$ and breaking it into two components: one for the specular reflection as given earlier, and one for the diffuse scattering. The diffuse scattering can be calculated from:

$$L_{SCAT1dl}(t) = \begin{bmatrix} k_{dVVl} & k_{dVHl} \\ k_{dHVl} & k_{dHHl} \end{bmatrix} \left(\frac{\lambda}{(4\pi)^{3/2}} \right) \left(\frac{1}{r_{1l}r_{2l}} \right) e^{-j\beta(r_{1l}+r_{2l})} e^{-j\beta v_m t \cos\varphi_{Rx}} \tag{4.57}$$

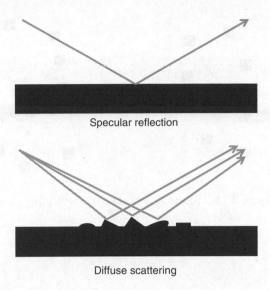

Specular reflection

Diffuse scattering

Figure 4.11   Comparison of a single specular reflection and diffuse scattering.

In this instance, there are different polarised reflections, k_{dVVl}, k_{dVHl}, k_{dHVl} and k_{dHHl}. It is interesting to point out that the power loss law is calculated differently for the two types of scattering. In the general case, the resultant impulse response is a summation of the N_{rs} specular and N_{ds} diffuse scatterers. If there is a LOS component also then Equation (4.56) needs to be added as well, which is assumed not to be time variant if the distance the mobile is moving is small.

$$\mathbf{h}(t, \tau) = \sum_{l=1}^{N_{rs}} L_{\mathrm{SCAT1r}l}(t)\delta(\tau - \tau_{\mathrm{SCAT1r}l}) + \sum_{l=1}^{N_{ds}} L_{\mathrm{SCAT1d}l}(t)\delta(\tau - \tau_{\mathrm{SCAT1d}l}) \qquad (4.58)$$

4. Double scattering

The original geometrically-based stochastic models assumed that the base station was placed high above the surrounding buildings and therefore there were no scatterers in its vicinity. On the contrary, there were scatterers around the mobile station. As the models evolved to address situations where the base station was also placed at lower heights, it became obvious to assume that there were also scatterers around that end of the link too. This gives rise to the double scattering case shown in Figure 4.12.

Double scattering has a significant impact on the radio system performance because it changes the statistical characteristics of the channel relative to the single scattering scenario. Figure 4.13 compares the cumulative distribution of a single Rayleigh and double Rayleigh channel. A double Rayleigh case is defined as the product of two Rayleigh channels and it can be seen that the double case is worse in the sense that for the same mean value, the outage value is lower in the double Rayleigh case for the same value of the outage probability.

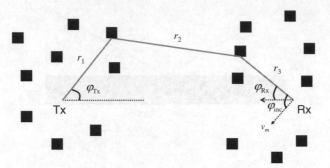

Figure 4.12   Illustration of a double scattering case.

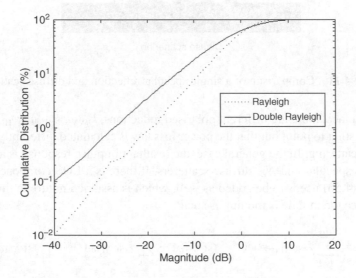

Figure 4.13   Comparison of a Rayleigh and Double Rayleigh SISO channel distribution.

5. The keyhole effect

   Another very important situation that arises in the double scattering scenario is the 'keyhole effect' (also known as the 'pinhole effect') (Chizhik, Foschini and Valenzuela, 2000; Chizhiz *et al.*, 2002; Gesbert *et al.*, 2002). This is illustrated in Figure 4.14. There are a lot of scatterers around the transmitting end, and a lot of scatterers around the receiving end. However all the waves from the scatterers at one one end have to pass through a common point to reach the other end, that is, a keyhole. This introduces a co-dependence of the links, although locally at one end or the other they might appear independent. The resulting channel capacity is low.

   Strictly speaking, there are certain propagation mechanisms that can give rise to a keyhole phenomenon, for example diffraction over building edges. However, the

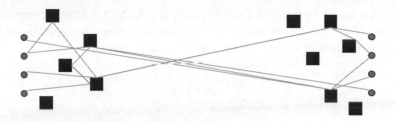

Figure 4.14   Illustration of the keyhole effect in a MIMO channel.

keyhole effect is more of a theoretical construct than an actually observed propagation scenario. In most situations, there is a richness of ways for the signal from the
Tx to reach the Rx and it is unlikely that all of them have to go through the same
keyhole point.

4.2.2.2   Parametric Stochastic Models

The second type of stochastic based models that we will discuss is that of parametric
stochastic models, and we will illustrate it through the dual directional channel model
(Steinbauer *et al.* 2001, Xu *et al.*, 2004). The description of time invariant SISO channels
was a function of the delay τ, that is, it reflected how much power arrives at different
delays. Clearly this power is arriving at the receiver from a certain angle that is referred to
as Angle of Arrival (AoA) φ_{Rx} and is associated with a certain Angle of Departure (AoD)
φ_{Tx} at the transmitter side as in Figure 4.15. Therefore a more detailed description of a
SISO channel impulse response with N_p taps at delays τ_l would be a function of the form:

$$h(\tau, \varphi_{Tx}, \varphi_{Rx}) = \sum_{l=1}^{N_p} a_l \delta(\tau - \tau_l) \sqrt{p'(\varphi_{Tx,l}, \varphi_{Rx,l})} e^{-j\vartheta_r} \qquad (4.59)$$

The function $p'(\varphi_{Tx,l}, \varphi_{Rx,l})$ is called the normalised power azimuth spectrum and incorporates both the angle of departure and the angle of arrival (in the earlier section we
defined it only as a function of one of them). The average tap power $E[|a_l|^2]$ is specified by
the power delay profile, and the taps are assumed to follow known statistics (e.g. Rayleigh).
The random phase ϑ_r is uniformly distributed for Rayleigh distributed channel taps.

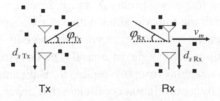

Figure 4.15   Geometry for dual directional channel model.

The model can now be extended to be used for the MIMO case by generating for the channel between the mth Tx antenna and n^{th} Rx antenna. Assuming uniform linear arrays at both ends with element separation d_{Tx} and d_{Rx} respectively:

$$h_{mn}(\tau, \varphi_{Tx}, \varphi_{Rx}) = \sum_{l=1}^{N_p} a_l \delta(\tau - \tau_l) \sqrt{p'(\varphi_{Tx,l}, \varphi_{Rx,l})} e^{-j\vartheta_r}$$

$$\times\ e^{-j\beta(m-1)d_{Tx}\sin\varphi_{Tx}} e^{-j\beta(n-1)d_{Rx}\sin\varphi_{Rx}} \tag{4.60}$$

The implicit assumption is again that the scatterers are a far distance away from the antennas and that the AoAs and AoDs correspond to plane waves at either end of the communication link, which in turn implies that the phase difference between elements depends on the relative orientation of the elements with respect to the plane wave angle in Figure 4.15.

Finally, consideration can be given to the fact that the mobile moves so the channel changes over time t. This works in a similar way to the previous model if the direction of motion at velocity v_m is as specified in Figure 4.15.

$$h_{mn}(\tau, \varphi_{Tx}, \varphi_{Rx}, t) = h_{mn}(\tau, \varphi_{Tx}, \varphi_{Rx})e^{-j\beta v_m t \cos\varphi_{Rx}} \tag{4.61}$$

Such a model has accuracy in terms of defining a wideband channel, although it requires extensive amounts of data dependent on four variables, which creates high complexity and also difficulty in generating the output of such data. If polarisation is incorporated also into the model, it creates further demands on data storage. This has been one of the main motives to produce more simplified models such as those described in earlier sections of this chapter.

4.2.2.3 Correlation-based Models

The analysis in Section 4.1.2 showed that the channel gains at two neighbouring points A and B within the same local area are two random variables that follow the same statistics and have a certain correlation. Specifically, if their envelopes follow Rayleigh statistics (i.e. the real and the imaginary parts of the complex channel gain are zero-mean, Gaussian random variables of the same variance), then the two random variables are related by a correlation coefficient ρ that depends on the angular spread of the received signal within this local area. Therefore the channel coefficients that would have been observed by two receive antennas placed at these two points A and B from a common transmit antenna would be complex Gaussian random variables with a correlation coefficient that depends on the power azimuth spectrum at the receiver, which is sometimes referred to as the power azimuth spectrum of the direction (angle) of arrival (DoA or AoA respectively).

By reciprocity, the channel gains from two remote transmit antennas to the same receive antenna would be related by a correlation coefficient that depends on the power azimuth around the transmitter array location, which is sometimes referred to as power azimuth spectrum of the direction (angle) of departure (DoD or AoD respectively). Correlation based models abstract from the actual power azimuth spectra at the receive and transmit sides and the underlying distribution of scattering objects that caused them. They are based solely on the correlations between channel coefficients.

We are interested in the channel transfer matrix H, which is constructed such that the element h_{mn} is the channel coefficient between transmitter m and receiver n. Let us define the *complex* correlation ρ_{ij} between two channel coefficients at the receive end, which we can derive as h_{mi} and h_{mj}, thus Tx branch m is fixed. The channel coefficients are derived in baseband and therefore they are complex quantities. Moreover we look at the narrowband situation first:

$$\rho_{ij,\text{Tx}m}^{\text{Rx}} = \frac{E[h_{mi}h_{mj}^*] - E[h_{mi}]E[h_{mj}^*]}{\sqrt{\text{var}[h_{mi}]\text{var}[h_{mj}]}} \tag{4.62}$$

where the notation E[] is the expected or average of the discrete samples and var[] is their variance. Assuming that the signals are zero-mean random variables, and that they have unit variance, the above expression simplifies to:

$$\rho_{ij,\text{Tx}m}^{\text{Rx}} = E[h_{mi}h_{mj}^*] \tag{4.63}$$

The correlation between two transmit branches, h_{in} and h_{jn}, can be therefore defined as follows for fixed receiver branch n:

$$\rho_{ij,\text{Rx}n}^{\text{Tx}} = E[h_{in}h_{jn}^*] \tag{4.64}$$

In order to simplify the notation, we introduce the vector H_{vec} which is constructed by stacking the columns of H, one below the other. The channel matrix H for a system with N_{Tx} transmit antennas and N_{Rx} receive antennas has dimensions $N_{\text{Rx}} \times N_{\text{Rx}}$, and therefore the vector H_{vec} has dimensions $(N_{\text{Tx}}N_{\text{Rx}}) \times 1$. The correlation matrix R which captures all the correlation coefficients $\rho_{ij,\text{Tx}m}^{\text{Rx}}$ and $\rho_{ij,\text{Rx}n}^{\text{Tx}}$ is defined as:

$$R_{\text{MIMO}} = E\left[H_{\text{vec}}H_{\text{vec}}^H\right] \tag{4.65}$$

R is a positive definite and complex conjugate symmetric. The superscript ^{H} is termed as the hermitian transpose, that is, the matrix is first transposed then the conjugate function is applied to each element. This is the quantity that correlation based models are interested in because it captures the co-dependence for every combination of transmit-receive antennas. If the correlation matrix is known and the channel statistics are complex Gaussian, then one can easily generate the channel matrices.

The simplest case of a correlation based model is the following: the elements of the channel transfer matrix are independent, identically, Rayleigh (complex Gaussian) distributed random variables. In MATLAB the relevant command to generate one realisation of a matrix would be

```
H =   (randn(nrx, ntx) + j * randn(nrx, ntx))/sqrt(2);
```

Clearly, repeating the command would generate more realisations, independent of each other.

Similarly to the approach we followed to generate samples of a channel that were correlated in time, we can use the correlation matrix to generate samples that are correlated in space. We generate a vector **h** of dimensions $N_{Tx} \times N_{Rx}$ with elements that are independent, complex Gaussian, circularly symmetric random variables.

```
h = (randn(nrx * ntx, 1) + j * randn(nrx * ntx, 1))/sqrt(2);
```

We define the matrix $W = R^{1/2}$, that is, a matrix such that $WW^H = R$. The matrix W can be created in MATLAB in different ways:

▶ Using the command sqrtm:

```
W =  sqrtm(R)
```

This approach gives the principal square root, that is, the unique matrix that satisfies $WW^H = R$ and whose eigenvalues has positive real parts.

▶ Using the command chol:

```
W = chol(R)
```

This approach gives the Cholesky factorisation of R, that is, a matrix that is triangular.[15]

We then create the vector $w = Wh$:

```
w = W * h;
```

---

[15] There are two ways to perform that Cholesky factorisation of a matrix A that is assumed to be hermitian (complex conjugate symmetric).

`R = chol(A); or L = chol(A,'lower');`.

The first command produces an upper triangular matrix R such that $R^H * R = A$. The second command produces a lower triangular matrix L such that $L * L^H = A$. If the original matrix A is not positive definite, the appropriate way to perform the Cholesky factorisation would be by calling one of the following commands:

`[R,p] = chol(A); or [L,p] = chol(A,'lower');`.

If A is positive definite, then $p = 0$ and R, L are the same as above. But if A is not positive definite, then p is a positive integer and when A is full, e.g. R is an upper triangular matrix of order $q = p - 1$ such that that $R^H * R = A(1:q,1:q)$. When A is sparse, R is an upper triangular matrix of size $q \times n$ so that the L-shaped region of the first q rows and first q columns of $R^H * R$ agree with those of A.

It can easily be shown that the elements of **w** are Gaussian circularly symmetric distributed (and therefore have Rayleigh distributed envelope), and have the desired correlation properties. From **w** we can construct the channel transfer matrix, **H** by appropriate reshaping:

```
H = reshape(w, nrx, ntx);
```

Given the general definition of correlation based models, we will look at more detail into two special cases of correlation-based models, namely the Kronecker model and the Weichselberger model. The contribution of these two models is that they introduce some structure to the correlation matrix.

The Kronecker model as illustrated in Figure 4.16 derives its name from the Kronecker matrix multiplication. As a reminder, let us assume that we have two matrices A and B of dimensions $N \times M$ and $K \times L$ respectively. We denote as a_{ij} the item that us on the ith row and j^{th} column of the matrix A. The Kronecker product $A \otimes B$ of A and B is defined as:

$$A \otimes B = \begin{bmatrix} a_{11}B & a_{12}B & \cdots & a_{1M}B \\ a_{21}B & a_{22}B & \cdots & a_{2M}B \\ \cdots & \cdots & \cdots & \cdots \\ a_{N1}B & a_{N2}B & \cdots & a_{NM}B \end{bmatrix} \tag{4.66}$$

There are two fundamental assumptions underlaying the Kronecker model:

▶ The correlations are the same independently of where they are calculated. The correlation of the links from two transmit antennas to a single receive antenna is the same, independently of which receive antenna is used for the calculation. Similarly for the links from any transmit antenna to two given receive antennas so it does not matter which transmit antenna we use. For example in the case of a 2×2 MIMO link, we can define our transmit correlation between the two transmit branches 1 and 2 in the following ways:

$$\rho_{12,Rx1}^{Tx} = \rho_{12,Rx2}^{Tx} = \rho_{21,Rx1}^{Tx} = \rho_{21,Rx2}^{Tx} \tag{4.67}$$

Given that the independence of which Rx branch is used, the transmit correlation can be simplified as follows:

$$\rho_{12}^{Tx} = \rho_{21}^{Tx} \tag{4.68}$$

Likewise for the two receive branches 1 and 2, we can define the following receive correlations:

$$\rho_{12,Tx1}^{Rx} = \rho_{12,Tx2}^{Rx} = \rho_{21,Tx1}^{Rx} = \rho_{21,Tx2}^{Rx} \tag{4.69}$$

> ## There are two fundamental assumptions underlaying the Kronecker model: (Continued)

Therefore:

$$\rho_{12}^{Rx} = \rho_{21}^{Rx} \qquad (4.70)$$

Having now made these simplifications, the correlation between two receive branches i and j in any MIMO channel can therefore be defined as ρ_{ij}^{Rx} and at any two transmit branches as ρ_{ij}^{Tx}. Therefore within any N × M MIMO matrix, the transmit correlation matrix $\boldsymbol{R}_{Tx}$ and the receive correlation matrix $\boldsymbol{R}_{Rx}$ are defined such that their elements ρ_{ij}^{Tx} and ρ_{ij}^{Rx} can therefore be defined for fixed arbitrary Tx branch m and fixed arbitrary Rx branch n as follows:

$$\rho_{ij}^{Rx} = E[h_{mi}h_{mj}^*] \qquad (4.71)$$

$$\rho_{ij}^{Tx} = E[h_{in}h_{jn}^*] \qquad (4.72)$$

Therefore $r_{Tx,ij}$ is the element on the ith row and j^{th} column of the matrix $\boldsymbol{R}_{Tx}$ and corresponds to the correlation between the signals from transmit antenna i and transmit antenna j. Note that the matrix $\boldsymbol{R}_{Tx}$ has dimensions $N_{Tx} \times N_{Tx}$, where N_{Tx} is the number of transmit antennas, so as to capture the correlation between any two of them. Similarly $r_{Rx,ij}$ is the element that is on the ith row and j^{th} column of the matrix $\boldsymbol{R}_{Rx}$ and corresponds to the correlation between the signals at receive antenna i and receive antenna j. The matrix $\boldsymbol{R}_{Tx}$ has dimensions $N_{Rx} \times N_{Rx}$, where N_{Rx} is the number of receive antennas.

▶ The correlations are separable

The correlation between two links can be split into the product of two correlation values: the correlation at the transmit side (which involves only the transmit antennas of the two links), and the correlation at the receive side (which involves only the receive antennas of the two links). Mathematically, this can be written for the correlation between the coefficient for transmit branch i to receive branch j, h_{ij} and the coefficient for transmit branch k to receive branch l, h_{kl} so that the following relation holds:

$$\rho_{ij,kl} = E[h_{ij}h_{kl}^*] = \rho_{ik}^{Tx}\rho_{jl}^{Rx} \qquad (4.73)$$

Implicitly, this means that the scattering situation around the transmitter and around the receiver, and hence also the corresponding correlation values, are separable from each other. From a physical standpoint, the Kronecker model describes a situation where there are several scatterers between the transmitter and receiver and the behaviour around the transmitter has no relation to the behaviour around the receiver.

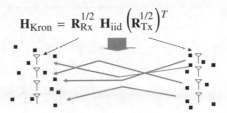

$$\mathbf{H}_{\text{Kron}} = \mathbf{R}_{\text{Rx}}^{1/2} \, \mathbf{H}_{\text{iid}} \left(\mathbf{R}_{\text{Tx}}^{1/2}\right)^T$$

Figure 4.16 Illustration of the basis behind the Kronecker MIMO channel model.

The Kronecker model therefore introduces some structure to the correlation matrix and states that:

$$\mathbf{R}_{\text{MIMO}} = \mathbf{R}_{\text{Tx}} \otimes \mathbf{R}_{\text{Rx}} \tag{4.74}$$

that is, the correlation matrix $\mathbf{R}_{\text{MIMO}}$ is given as the Kronecker product of the correlation matrices, $\mathbf{R}_{\text{Tx}}$ at the transmitter and $\mathbf{R}_{\text{Rx}}$ at the receiver (Kermoal *et al.* 2002).

Example: In the case of a 2×2 channel, the channel transfer matrix is of dimension 2×2, and therefore has 4 elements corresponding to each one of the transmit-receive links. Therefore the correlation matrix $\mathbf{R}$ is a 4×4 matrix. The transmit correlation matrix $\mathbf{R}_{\text{Tx}}$ is a 2×2 matrix. The receive correlation matrix $\mathbf{R}_{\text{Rx}}$ is a 2×2 matrix.

$$\mathbf{R}_{\text{Tx}} = \begin{bmatrix} \rho_{11}^{\text{Tx}} & \rho_{21}^{\text{Tx}} \\ \rho_{12}^{\text{Tx}} & \rho_{22}^{\text{Tx}} \end{bmatrix} \quad \mathbf{R}_{\text{Rx}} = \begin{bmatrix} \rho_{11}^{\text{Rx}} & \rho_{21}^{\text{Rx}} \\ \rho_{12}^{\text{Rx}} & \rho_{22}^{\text{Rx}} \end{bmatrix} \tag{4.75}$$

$$
\begin{aligned}
\mathbf{R}_{\text{MIMO}} &= \begin{bmatrix} \rho_{11}^{\text{Tx}} \mathbf{R}_{\text{Rx}} & \rho_{21}^{\text{Tx}} \mathbf{R}_{\text{Rx}} \\ \rho_{12}^{\text{Tx}} \mathbf{R}_{\text{Rx}} & \rho_{22}^{\text{Tx}} \mathbf{R}_{\text{Rx}} \end{bmatrix} \\[2mm]
&= \begin{bmatrix} \rho_{11}^{\text{Tx}} \rho_{11}^{\text{Rx}} & \rho_{11}^{\text{Tx}} \rho_{21}^{\text{Rx}} & \rho_{21}^{\text{Tx}} \rho_{11}^{\text{Rx}} & \rho_{21}^{\text{Tx}} \rho_{21}^{\text{Rx}} \\ \rho_{11}^{\text{Tx}} \rho_{12}^{\text{Rx}} & \rho_{11}^{\text{Tx}} \rho_{22}^{\text{Rx}} & \rho_{21}^{\text{Tx}} \rho_{12}^{\text{Rx}} & \rho_{21}^{\text{Tx}} \rho_{22}^{\text{Rx}} \\ \rho_{12}^{\text{Tx}} \rho_{11}^{\text{Rx}} & \rho_{12}^{\text{Tx}} \rho_{21}^{\text{Rx}} & \rho_{22}^{\text{Tx}} \rho_{11}^{\text{Rx}} & \rho_{22}^{\text{Tx}} \rho_{21}^{\text{Rx}} \\ \rho_{12}^{\text{Tx}} \rho_{12}^{\text{Rx}} & \rho_{12}^{\text{Tx}} \rho_{22}^{\text{Rx}} & \rho_{22}^{\text{Tx}} \rho_{12}^{\text{Rx}} & \rho_{22}^{\text{Tx}} \rho_{22}^{\text{Rx}} \end{bmatrix}
\end{aligned} \tag{4.76}
$$

Model validation

One of the ways to test the validity of a MIMO channel model is by comparing the cumulative distribution functions of the eigenvalues of the *measured* channel matrices and those of the eigenvalues of *synthetic* channel matrices that have been generated according to the model in question.

Let us look for example at Figure 4.17, which takes actual measurements of a channel measured in an indoor environment and compares with the Kronecker model to test its validity. The parameters of the Kronecker model, that is, the transmit and receive correlations, have been calculated based on measurements of the 4×4 system, synthetic channels

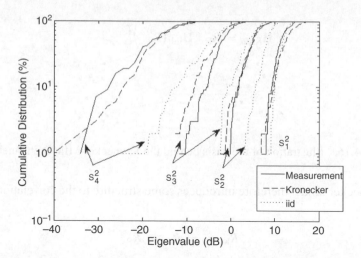

Figure 4.17 Comparison of the eigenvalue distributions of a channel using the Kronecker model compared to an iid model and NLOS measurement data.

have been generated based on the Kronecker model and the distribution of the eigenvalues of the resulting simulated channel transfer matrices are shown in the figure. Clearly, the simulated eigenvalue distributions are close to the actual ones. Figure 4.17 also shows the eigenvalue distributions for an i.i.d. case. The second, third and fourth eigenvalue CDFs are significantly offset relative to those of the measurement and Kronecker model results. Therefore, the lower values of the measured/ synthetic eigenvalues indicate that the capacity of this system is lower than that of a channel with i.i.d terms. Moreover, the steepness of the CDFs indicates that the diversity order is lower than that of an i.i.d. case. Therefore, correlation limits both capacity and diversity order.

If we apply the Kronecker principle, the generation of correlated channel entries is largely simplified. Instead of dealing with a large matrix of dimension $(N_{Tx} N_{Rx}) \times (N_{Tx} N_{Rx})$, we can deal with a much smaller matrix of dimension $(N_{Tx} \times N_{Rx})$

$$H_{Kron} = R_{Rx}^{1/2} H_{iid} \left(R_{Tx}^{1/2} \right)^{T} \tag{4.77}$$

It can easily be verified that the coefficients that are thus derived have the desired correlation properties. Again $R^{1/2}$ is the matrix such that $R^{1/2} \left(R^{1/2} \right)^{T} = R$ and can be found for example from the Cholesky factorisation of a correlation matrix.

The Kronecker model is therefore a simple correlation based stochastic model to implement with few driving parameters, namely the signal correlations at each end.

The Weichselberger Model is an extended version of the Kronecker model, which is not limited by the assumption of separability (Weichselberger *et al.* 2004). The first step is to decompose the correlation matrices at the transmitter and receiver ends by eigenvalue decomposition as follows:[16]

$$R_{Rx} = U_{Rx} \Lambda_{Rx} U_{Rx}^H \qquad (4.78)$$

$$R_{Tx} = U_{Tx} \Lambda_{Tx} U_{Tx}^H \qquad (4.79)$$

where Λ_{Tx} and Λ_{Rx} are eigenvalue matrices with corresponding orthonormal eigenvectors, U_{Tx} and U_{Rx}. The matrices $R_{Rx}^{1/2}$ and $R_{Tx}^{1/2}$ as mentioned earlier are not unique and therefore we can choose:

$$R_{Rx}^{1/2} = U_{Rx} \sqrt{\Lambda_{Rx}} \qquad (4.80)$$

$$R_{Tx}^{1/2T} = \sqrt{\Lambda_{Tx}} U_{Tx}^H \qquad (4.81)$$

where

$$\sqrt{\Lambda_{Rx}} = \mathrm{diag}(\lambda_{Rx,1}, \lambda_{Rx,2}, ...\lambda_{Rx,\tilde{n}x}), \qquad (4.82)$$

$$\sqrt{\Lambda_{Tx}} = \mathrm{diag}(\lambda_{Tx,1}, \lambda_{Tx,2}, ...\lambda_{Tx,\tilde{n}x}). \qquad (4.83)$$

We substitute these into Equation (4.77) and we obtain

$$H_{\text{Weich}} = U_{Rx} \sqrt{\Lambda_{Rx}} H_{\text{iid}} \sqrt{\Lambda_{Tx}} U_{Tx}^H \qquad (4.84)$$

Let us also define the auxiliary matrix Ω_w:

$$\Omega_w = \begin{pmatrix} \lambda_{Rx,1} & & & \\ & \lambda_{Rx,2} & & \\ & & \cdot & \\ & & & \cdot \\ & & & & \cdot \\ & & & & & \lambda_{Rx,N\tilde{n}x} \end{pmatrix} \left(\lambda_{Tx,1} \lambda_{Tx,2} ....\lambda_{Tx,N\tilde{n}x} \right) \qquad (4.85)$$

---

[16] Because correlation matrices are complex conjugate symmetric, they have the same left and right eigenvectors.

The contribution of the model comes from noticing that Equation (4.85) can be written as:

$$H_{\text{Weich}} = U_{\text{Rx}} \mathbf{\Omega}_{\text{w}} \odot H_{\text{iid}} U_{\text{Tx}}^H \qquad (4.86)$$

where $\odot$ represents an element-wise multiplication of the matrices. If the matrix $\mathbf{\Omega}_{\text{w}}$ has the form of Equation (4.85), then the model reduces to the Kronecker model. In the general case, the matrix $\mathbf{\Omega}_{\text{w}}$ can have an arbitrary structure and is the driving parameter for the model.

Figure 4.18 gives example cases of how the matrix can be used to model different non separable channel scenarios. In each of the four sub-figures (a) to (d), an asterisk (*) is used to represent where the elements are more significant in value and the elements with a zero are negligible in value.

In Figure 4.18 (a) the top row and left column are more significant. This case would result in a scenario where there are many scatterers at each end and the channel will be separable, so the result is close to that of the Kronecker model. If all the elements were comparable to each other (i.e. a matrix full of asterisks) then the channel would be similar to that of an i.i.d case which can be seen by inspection of Equation (4.86). In such a case there would be scatterers everywhere.

In contrast, Figure 4.18 (b) shows a case where $\mathbf{\Omega}_{\text{w}}$ is like a diagonal matrix. This means essentially that each output eigenvector (column of the matrix U_{Rx}) is coupled to only one of the input eigenvectors (column of the matrix U_{Tx}). The Tx and Rx are also clearly linked so this channel is definitely not separable.

Figure 4.18 (c) shows an instance where the left column is more significant. This causes there to be more scattering around the Rx so the Tx is a more correlated environment with

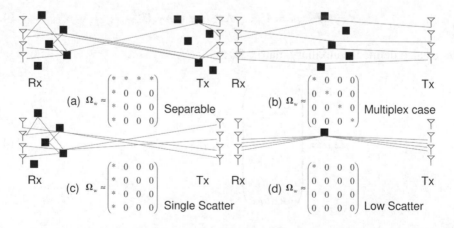

Figure 4.18 Illustration of the parameters applied in the Weichselberger model that can be applied to define physical channel states.

Table 4.1 Comparison of three stochastic model driving parmeters

Model name	Number of parameters used to drive the model
Kronecker	$M_{Tx}^2 + N_{Rx}^2$
Weichselberger	$M_{Tx}(M_{Tx} - 1) + N_{Rx}(N_{Rx} - 1) + M_{Tx}N_{Rx}$

less scattering. This will cause some coupling between the Tx and Rx as the angle of arrival and angle of departure from one end to the other will depend on each other. If the matrix had a more significant top row, it would cause the opposite effect where there would be no scattering around the Rx but much around the Tx.

Finally Figure 4.18 (d) is a case where there is very little scattering and all the paths are very similar to each other. In such a case there would not be much multiplexing potential.

The Kronecker model depends on two correlation matrices, which add slightly more complexity than the simple i.i.d channel though they are easier to evaluate. The Weichselberger model may improve accuracy significantly in some cases though at an increased cost of forming driving vectors at each end. Thus computation is increased as compared to Kroenecker in Table 4.1, though this model can allow for circumstances where the channel is not suitably separable.

Stochastic modelling

Unlike deterministic models, stochastic models are based on random processes using suitable input parameters. In this case they are not designed to match one specific environment with precise accuracy but they can match several radio environments with an acceptable accuracy, thus creating a more general model for several radio environments. The greatest challenge with stochastic models is producing them in a way in which they do provide acceptable accuracy while maintaining simplicity. If there are too many input parameters, then the complexity will increase to that of an equivalent deterministic model. Different forms of stochastic model are summarised as follows by controlling different parameters:

▶ Geometrically based stochastic models rely on placing scattering objects around the transmitter and the receiver at locations that are chosen randomly but follow known distributions. Therefore the number of scatterers and their random positions are the main parameters in such a model.

▶ Parametric stochastic models generate impulse responses with known power delay profiles and random delay tap realisations, where both the tap amplitude and the angular properties follow known distributions. Therefore the main input parameters in this case are the random delay tap values, their magnitude and their angle of arrival or departure.

Stochastic modelling (Continued)

▶ Correlation based models generate channel impulse responses with known power delay profiles and random tap amplitudes, such that the tap amplitudes follow known distributions and have known autocorrelation properties. It is also the case that narrowband fading channels can be generated directly from correlation based models as well as a wideband impulse. Therefore in such models, the driving parameter is the correlations between the branches and less attention is given to the physical structure of the channel.

In the above three cases of stochastic models, they are generated as random processes while going down the list, there is less and less information given about the physical structure of the channel. This will reduce the complexity of the channel while also beginning to compromise accuracy, though at the same time moving towards a more general model.

4.3 Summary

The following table summarises the deterministic models that have been introduced in this chapter, which will be of benefit with regard to detailed accuracy and point to point comparisons where computation time is not an expense. These models have been used to illustrate how the interdependence between MIMO branches is accounted for in fine detail.

Model	Method	Application
Recorded impulses	Takes direct measurements of wideband impulse responses to be stored and applied as a model.	Would normally be applied to a very specific environment where a theoretically generated model is not suitably accurate. Often a last resort due to high cost of producing such results. More used to validate other models.
Ray tracing	Calculate the angles of arrival and departure of the rays generated within a channel and determine their magnitude based on the electromagnetic properties of the scattering objects in the environment.	Large quantities of data about the environment are required, which are often frequency dependent. For large environments the data can be too overwhelming. Thus ray tracing is often only suited to small indoor environments or other locations where there are few scatterers.

Finally the table on the next page summarises the stochastic models that have been introduced in this chapter, which will be of benefit to applications requiring simplicity in implementation, a more general model of several scenarios but a compromise in accuracy when being used to test a system for a specific scenario. The interdependence between the MIMO branches is still considered between the models, though some specific scenarios that occur in MIMO channels may be omitted depending on the simplicity.

Model	Method	Application
Geometric	Generation of random scatterers between a transmitter and receiver from which the angles of departure at the Tx and angles of departure at the Rx are determined along with the respective path delays, phase and magnitudes.	In cases where scatterers are specifically placed in locations that would cause special cases such as a "keyhole" effect or specific "clusters" for wideband channels that will occur in specific radio environments (e.g. suburban, urban, indoor).
Parametric	Direct generation of impulse responses with defined delay taps, where the angles of departure at the Tx and the angles of arrival at the Rx are input to determine the corresponding magnitudes and phases of the delay taps.	For cases where little is known about the structure of the scattering environment so that angles of departure and arrival as well as tap delays can be used instead as the driving parameters.
Weichselberger	Inputs a matrix that can be interpreted to define the rough position of the scatterers in a radio environment, which will in turn affect the interdependence of the first independent MIMO branches. No other physical information is considered.	For cases where very abstract information about the environment is known and environment scenarios can be easily changed so as to gain a more general idea of how a MIMO system will work in different environments.
Kronecker	Based on correlation at the Tx and Rx to then cause the independent MIMO branches to have some interdependence.	For cases where it can be assumed the scattering at the Tx and the Rx are separable, that is, have no dependence on each other, so that correlations can be used as simple input parameters.

5

MIMO Antenna Design

Tim Brown

The antenna is important to any wireless communication system in order to enable the transition from electromagnetic waves to an induced alternating current in the transmission line or the transition from alternating current to electromagnetic waves. In order for this to happen, there are two fundamental factors to antenna design; the first is the fact that antennas will radiate electromagnetic waves from different directions while a corresponding receive antenna will receive signals only from certain directions. This will affect how much of the transmitted power is received as well as the fact that it will decay due to path loss. The impact of the antennas is measured by what is known as *antenna gain* in a given direction or directions. Secondly we have to consider how reliably current (and likewise power) transfers from the transmission line to the antenna, or how power output from an antenna transfers into the transmission line. This is optimised by what is known as *impedance matching*. The biggest problem in antenna design is that it is difficult to optimise the impedance matching while also optimising antenna gain. Furthermore, in MIMO we have to build arrays of antenna elements, which means that for each element the antenna gain and the impedance matching will be affected in the presence of its neighbouring elements. For small MIMO antennas, there is a further problem in that when antenna elements have to be so close to each other, power will couple from one antenna to the other, which is due to what is known as *mutual impedance*. This mutual impedance can be beneficial in terms of reducing correlation between antenna branches which is beneficial to MIMO though at the same time gain can be reduced, which is not beneficial to MIMO.

This chapter will open up by summarising the fundamentals of antennas so that the importance of antenna gain and impedance matching can be properly understood. A number of concepts are related to impedance matching or antenna gain such as directivity, efficiency, polarisation, reciprocity and radiation resistance, which will be introduced in

Practical Guide to the MIMO Radio Channel with MATLAB® Examples, First Edition.
Tim Brown, Elisabeth De Carvalho and Persefoni Kyritsi.
© 2012 John Wiley & Sons, Ltd. Published 2012 by John Wiley & Sons, Ltd.

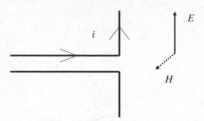

Figure 5.1   Illustration of a basic antenna structure.

this chapter. For more detailed explanation and analysis of these concepts, the reader is advised to consult other books such as Balanis (2005), Huang and Boyle (2008) and Saunders and Aragon-Zavala (2007). The remainder of the chapter will then consider some of the fundamental aspects affecting compact MIMO antenna design and the issues associated with mutual impedance.

5.1   Antenna Element Fundamentals

Before coming to the fundamentals of antenna elements, the basic function of an antenna needs to be understood. Its purpose is to make a transition from alternating current in a transmission line (e.g. a cable) to electromagnetic fields in the air as illustrated in Figure 5.1. Due to rules of *Maxwell's Equations* (Saunders and Aragon-Zavala 2007), an antenna will radiate an electromagnetic wave such that whenever it radiates an electric field or E-field in one direction, it will always radiate a magnetic field or H-field that is orthogonal to that E-field. Due to these orthogonal E-fields and H-fields there will be a resultant power density, which will be high in some directions while low in others depending on how strong the E-field is in that direction.

5.1.1   Isotropic Radiator

A perfect antenna could be considered as an isotropic radiator, which we could represent as a point source. If it is radiating isotropically then it radiates its power uniformly in all directions. Therefore, if a sphere has radius r, the surface area is $4\pi r^2$ so the power density, S_{iso}, (i.e. power per unit area) radiated at any point on the surface of the sphere can be expressed in terms of the source power, P_s, as:

$$S_{\mathrm{iso}} = \frac{P_s}{4\pi r^2} \tag{5.1}$$

which assumes that is r above a minimum value, which will be discussed later. Such an antenna is not possible to construct. An isotropic antenna, however, is used as a reference for quantifying other properties of a practical antenna.

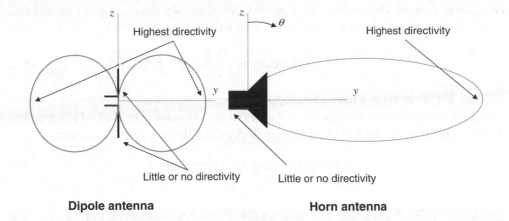

Dipole antenna   **Horn antenna**

Figure 5.2   Illustration of two antennas with low and high directivity.

5.1.2   Directivity and Gain

Since practical antennas are never isotropic, the directivity is defined as the ratio of the power density radiated at a point with a given angle and distance from an antenna compared to what would be radiated at that point by an isotropic antenna. In Figure 5.2 a comparison is shown between a low directivity or omnidirectional antenna known as a dipole and a high directivity horn antenna. Both antennas are assumed to be oriented in the yz plane, where directivity is represented as a polar plot which shows for different angles of θ (relative to the z axis), the directivity is variable. A dipole antenna will radiate most where θ is $\pm 90°$ and the least when θ is $0°$ and $180°$ as shown in Figure 5.2.

The horn antenna in Figure 5.2 shows a case where power density is radiated the strongest in the y direction so its directivity is plotted to be high in this direction, while at the rear of the antenna there is very little or no power density radiated so it has little directivity in this direction. The antenna is therefore considered to be directional because it radiates mostly in one direction when $\theta = 90°$. Clearly in both cases, the directivity can be used to measure how the antenna structure radiates power density in a certain direction or directions. The directivity, D, at angle, θ in this case can be defined when knowing the corresponding power density, $S_{\mathrm{dir}}(\theta)$ as:

$$D(\theta) = \frac{S_{\mathrm{dir}}(\theta)}{S_{\mathrm{iso}}} \qquad (5.2)$$

It can be seen from this equation that directivity is independent of r, because the ratio of the radiated power density to isotropic power density will not change. When defining the directivity, it is assumed that the materials used to build the antenna have no loss in them, which in practice is never true and so the actual power radiated will be less than $S_{\mathrm{dir}}(\theta)$. This is because some of the power going into an antenna will turn into heat and be lost, rather than making the transition from alternating voltage to electromagnetic fields.

This therefore means the antenna has a defined efficiency, which is a ratio of total radiated power (TRP) to what power is input to the antenna, P_s. Therefore:

$$\text{Efficiency} = \frac{\text{TRP}}{P_s} \tag{5.3}$$

Since antennas always have an inherent efficiency, the directivity is therefore rarely talked about in practice and likewise the gain is a quantity that can be applied to a practical situation. Gain is also related to the directivity by the following simple equation:

$$\text{Gain} = \text{Efficiency} \times \text{Directivity} \tag{5.4}$$

Gain and efficiency

Two important parameters which are talked about in antenna design are that of gain and efficiency. In order to understand gain and efficiency, it is important to understand the concept of an isotropic antenna, as well as a practical antenna design that has a directivity. The four terms are explained as follows:

▶ No antenna radiates power density spherically in all directions, thus no antenna can be perfectly isotropic. For an isotropic antenna, the power density radiated, S_{iso}, is defined as the source power, P_s divided by the area of a sphere at radius r as follows:

$$S_{iso} = \frac{P_s}{4\pi r^2} \tag{5.5}$$

▶ Directivity is defined as the ratio of the power density radiated or received by an antenna in a certain direction compared to what would be radiated or received by an isotropic antenna in the same direction.
▶ Any practical antenna will not radiate all the source power, P_s and some will be lost as heat. Therefore the ratio of the total power radiated compared to the source power is defined as the efficiency.
▶ The gain is the same principle as directivity but efficiency is also taken into account. Thus the two are related by the simple equation:

$$\text{Gain} = \text{Directivity} \times \text{Efficiency} \tag{5.6}$$

5.1.3 Far Field and Rayleigh Distance

In order for Equation (5.1) to be valid, it was stated that r is assumed to far enough away from the antenna and in this case it is known to be in the *far field*. When the radiated power

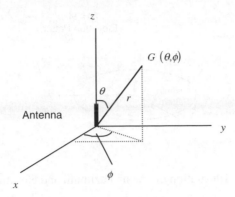

Figure 5.3 Illustration of antenna gain patterns in polar coordinates.

density is measured from any antenna in the far field, it will be proportional to $1/r^2$ and it will not be the case at distances lower than this. The minimum distance required to be in the far field, is defined by the *Rayleigh distance*, r_d of an antenna, which can be found by knowing the wavelength, λ and the maximum dimension of the antenna, D_m, as follows (Balanis 2005):

$$r_d = \frac{2D_m^2}{\lambda} \tag{5.7}$$

5.1.4   Three Dimensional Antenna Patterns

So far, the gain patterns of an antenna shown in Figure 5.2 are in two dimensions, requiring only one angle θ, while patterns can also be considered in three dimensions as shown in Figure 5.3. In this case, two spherical angles that rotate about the origin are shown, θ and ϕ, where gain, $G(\theta, \phi)$ is then defined for a fixed arbitrary value of distance r. It is also equally valid to define the directivity at any point in the same way as gain by using the notation $D(\theta, \phi)$, though this is not commonly used. For an isotropic antenna, both the gain and directivity always have a value of one no matter what the value of θ and ϕ. However, it is often useful to express the gain and directivity of an antenna in decibels. In this instance the gain and directivity of an isotropic antenna is 0 dBi. This means it is zero decibels compared to an isotropic antenna. In the general form therefore, the gain at any given value of θ and ϕ is defined as the power density received or transmitted at angles θ and ϕ compared to what would be received or transmitted from an isotropic antenna at the same arbitrary distance r as follows:

$$G(\theta, \phi) = \frac{S(\theta, \phi)}{S_{iso}} \tag{5.8}$$

Two commonly expressed terms about an antenna are its *azimuth* pattern and *elevation* pattern. If the angle θ in Figure 5.3 was set to 90° and measurements were taken for

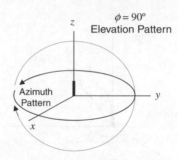

Figure 5.4   Illustration of antenna azimuth and elevation patterns.

$\phi = -180°$ to $180°$ then the gain values resolved versus ϕ will give the azimuth pattern. For an elevation pattern, ϕ could be fixed at either $0°$ or $90°$ and measurements then taken for $\theta = -180°$ to $180°$, which will give an elevation pattern. For clarity, however, it is usually appropriate to offset the elevation pattern by $90°$ since this would then mean that the same gain occurs at $0°$ in both the azimuth and elevation plots. The azimuth and $\phi = 90°$ elevation plots are therefore illustrated in Figure 5.4. It can be seen clearly that the gain values in the middle point, intersecting the positive y axis and the start/end points intersecting the negative y axis have the same gain values.

5.1.5   Impedance and Return Loss

As well as a good antenna having gain in a certain direction, its input must also have a suitable impedance that is matched to the source. Any antenna will have an input impedance, which can be defined as Z_{in} and this denotes the voltage to current ratio at a given frequency. It is known from classical circuit theory that with a matched impedance, the maximum amount of power can be transferred to the antenna. If there is a mismatch then some or all power will be reflected back into the source feed (in the case of transmit mode) or not sent from the antenna to the receiver (in the receive case). The standard formula to calculate the reflection coefficient (which is a ratio of voltage reflected back down a transmission line to that going into the antenna), Γ, is defined as:

$$\Gamma = \frac{Z_{in} - Z_0}{Z_{in} + Z_0} \tag{5.9}$$

where Z_0 is the source impedance, normally 50Ω. Using the reflection coefficient, the return loss is then normally defined in dB by $-20\log_{10}|\Gamma|$. Therefore, if the impedance is close to 50Ω, the return loss will be high. For a good antenna, return loss is expected to be at least 10 dB so that less than 10% of the power is reflected back down the transmission line if the antenna is in transmit mode. The return loss is usually measured to determine the antenna's frequency of operation and its bandwidth since in such a bandwidth we would normally see at least 10 dB return loss. Outside of this bandwidth, the return loss is usually small, close to 0 dB as would be the case for an open circuit when no antenna is present.

5.1.6   Reciprocity

Whether an antenna is a transmit (Tx) antenna or a receive (Rx) antenna the gain, impedance and field patterns all apply in the same way. Thus the antenna has a reciprocity whether it is in Tx or Rx mode. If two antennas are communicating with each other over a given distance, the resulting channel is the same whichever end a radio frequency is being transmitted from.

Antenna radiation patterns and antenna impedance

Antennas typically will have three principal radiation patterns (though there are some exceptions), where the most typical are defined as the *azimuth* and, $\phi = 0^o$ *elevation* and $\phi = 90^o$ *elevation* radiation patterns.

▶ The angle of elevation is the angle up or down from the horizontal plane, where an elevation pattern plots the antenna gain at a fixed azimuth angle ϕ as shown in Figure 5.3 from -180° below the horizontal plane, up to 180°. It should be noted that this is *not* the same as angle θ in Figure 5.3. which is relative to the vertical plane, the two angles are 90° apart. The azimuth angle, ϕ can be fixed at 0° or 90° to give the respective elevation patterns

▶ The azimuth pattern plots the gain versus azimuth angle ϕ around the antenna ranging from -180° up to 180°, while angle θ illustrated in Figure 5.3 is fixed to 90°.

If an antenna is highly directional and radiating in the vertical z direction as illustrated in Figure 5.3, then there is little value in characterising its azimuth pattern, since most of the power is radiating vertically. However, if the antenna was radiating in the y direction, then the azimuth and $\phi = 90^o$ elevation patterns would be of interest; the $\phi = 0^o$ elevation would be therefore negligible in comparison. Thus, in cases of directional antennas, not all three principal patterns are of interest, though for omni-directional antennas this is definitely the case.

Any antenna, no matter how high the gain is, must have a matched input impedance in order that power will transfer from the source, or to the receiver.

5.1.7   Antenna Polarisation

Up to this point, the gain of an antenna has been defined by the power density radiated in a given direction compared to an isotropic radiator but no consideration has been given to

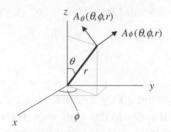

Figure 5.5   Illustration of the polarised E-field coming from an antenna.

the orientation of the E-field and consequently H-field defined by Maxwell's equations. If the E-field was vertical the H-field would be horizontal, though due to the way the antenna is built, it cannot be assumed to be always vertical, it could be tilted. For the purposes of this chapter, the E-fields radiated by an antenna will be represented as $E_\theta (\theta, \phi, r)$ and $E_\phi (\theta, \phi, r)$ as illustrated in Figure 5.5. These are not common notations to use but later on in this chapter they will help provide clarity in order to discern between E-fields radiated by an an antenna and E-fields incident on an antenna.

When the antenna is tilted, we have to define its tilt by separating it into two separate components, the vertical part, A_θ and the horizontal part, A_ϕ so that they add up by Pythagoras' rule where $|A| = \sqrt{|A_\theta| + |A_\phi|}$. If we consider an antenna radiating E-fields with different magnitudes with these vertical and horizontal components in all directions, then they can be represented in Figure 5.5 as fields dependent on ϕ, θ and r as shown. If we made θ equal to 90° and ϕ equal to 0°, the antenna could be oriented completely parallel to the z axis, in which case it would be completely vertically polarised in the x direction and be defined as $A_\theta (\theta, \phi, r)$. On the other hand, the antenna could be oriented parallel to the y axis in which case it would be horizontally polarised in the y direction and would be defined by $A_\phi (\theta, \phi, r)$. In practice, whatever angles and distance the power density is defined at, the E-fields radiated by the antenna will never be oriented perfectly horizontally or perfectly vertical due to imperfections in the antenna so they will therefore have a total radiated field, A defined by the following equation:

$$|A (\theta, \phi, r)| = \sqrt{|A_\theta (\theta, \phi, r)|^2 + |A_\phi (\theta, \phi, r)|^2} \tag{5.10}$$

Therefore the orientation of the total field at a fixed point defines its polarisation at that angle, which will have a combination of two E-fields which can be considered the vertical part and horizontal part, A_θ and A_ϕ, to define the polarisation. A high A_θ and low A_ϕ will define a predominantly vertical polarisation while the opposite will define a predominantly horizontal polarisation. These two separate fields also have subsequent power densities, S_θ and S_ϕ, which also means that they have polarised antenna gains, G_θ and G_ϕ. Having now established how polarisation is defined for the antenna, it should now be understood in terms of the radio environment. Figure 5.6 also shows an isotropic antenna where E-fields are incoming and they each have a polarisation at a given angle defined by $E_{i\theta}$ and $E_{i\phi}$.

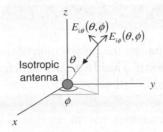

Figure 5.6   Illustration of the linear polarisations of the antenna.

These incident electric fields have been radiated from another transmitting antenna and they have propagated through the radio environment undergoing path loss, reflection, scattering etc. as discussed in Chapter 4. The captured electric field is transformed into a current on the receiving antenna itself according to the rules of reciprocity discussed in the previous section. This is because the ratio of the incoming field to the induced current when the antenna is acting as a receiver is equal to the ratio of the radiated electric field to the exciting current when the antenna is acting as a transmitter.

We will assume that the receiving antenna in this instance is isotropic as shown in Figure 5.6. We will assume that the transmitting antenna only radiates A_θ fields, though when they have propagated through the radio environment, scattering causes them to de-polarise so both $E_{i\theta}(\theta, \phi)$ and $E_{i\phi}(\theta, \phi)$ fields arrive at the receiving antenna.

If the two received fields are considered separately, a spherical integration can be performed to integrate the power of the two separate E-fields, which results in a vertical power, P_V and a horizontal power, P_H as follows:

$$P_V = \int_{-\pi}^{\pi} \int_0^{\pi} |E_{i\theta}(\theta, \phi)|^2 \sin\theta d\theta d\phi \tag{5.11}$$

$$P_H = \int_{-\pi}^{\pi} \int_0^{\pi} |E_{i\phi}(\theta, \phi)|^2 \sin\theta d\theta d\phi \tag{5.12}$$

The XPR is then defined as the ratio of the vertical to the horizontal power as follows:

$$\text{XPR} = \frac{P_V}{P_H} \tag{5.13}$$

We have also learned in previous sections that the antenna itself can have orthogonally polarised gains $G_\theta(\theta, \phi)$ and $G_\phi(\theta, \phi)$. These gains are important to distinguish since if the antenna has a G_θ pattern and no G_ϕ pattern then it is vertically polarised since it will only receive $E_{i\theta}$ fields. If the antenna has only a G_ϕ pattern it will only receive $E_{i\phi}$ fields. Antennas in practice have both G_θ and G_ϕ so they are never perfectly vertically or horizontally polarised, they will have a mixture of both vertical and horizontal polarisation.

Polarisation

The polarisation of the antenna and the polarisation of the radio environment are two important aspects of polarisation that need to be distinguished. In the case of the antenna, the polarisation can be defined as follows:

▶ The polarisation of the antenna concerns the electric fields radiated or received by an antenna, which have been denoted by the notations, A_θ and A_ϕ as the two polarisation components. These components occur at at a given point in space defined by angles θ and ϕ for a fixed distance in the far field of the antenna.
▶ Where an antenna has A_ϕ that are either zero or negligible compared to the A_θ components, it is considered as a vertically polarised antenna.
▶ Where an antenna has A_θ that are either zero or negligible compared to the A_ϕ components, it is considered as a horizontally polarised antenna.

For the case of the radio environment, the polarisation can be defined as follows:

▶ The vertical polarisation of the environment is a complete spherical integration of the power densities due to the incoming electric fields denoted as $E_{i\theta}$, resulting in a vertical received power, P_V.
▶ The horizontal polarisation of the environment is a complete spherical integration of the power densities due to the incoming electric fields denoted as $E_{i\phi}$ resulting in the horizontal received power, P_H.
▶ The cross polar ratio, XPR, is determined by the ratio of P_V to P_H.

5.1.8   Mean Effective Gain

The directivity and gain patterns applied so far are what would be found in a free space scenario with a single polarisation. In reality there are going to be objects around the antenna known as scatterers, which will result in the signals incoming to the receive antenna or outgoing to the transmit antenna being reflected multiple times. Depending on what angles these scatterers are at from the antenna, they will change the mean power received by the antenna and thus it is useful to define the mean effective gain (MEG) instead of free space gain that results from the antenna being in a real environment. By definition, the mean effective gain is the ratio of total power received by an antenna in a given radio environment and what power could be received by an isotropic antenna in that same environment. The total isotropic power is therefore equal to the sum of the total vertically polarised power, P_V and horizontal power P_H. The antenna itself will receive a proportion of P_V and P_H so that at a given angle the power received at angles θ and ϕ will be:

$$P_\theta(\theta, \phi) = P_V G_\theta(\theta, \phi) \qquad (5.14)$$

$$P_\phi(\theta, \phi) = P_H G_\phi(\theta, \phi) \qquad (5.15)$$

Thus P_θ is the power received due to gain G_θ at angles θ and ϕ, which can only receive vertically polarised power. P_ϕ is the power received due to gain G_ϕ at angles θ and ϕ, which can only receive horizontally polarised power. These powers will vary at each angle and as such the total vertically polarised power received by the antenna is a spherical integral of P_θ powers and the total horizontally polarised power received by the antenna is a spherical integral of P_ϕ powers as follows.

$$\text{MEG} = \frac{\int_{-\pi}^{\pi} \int_0^\pi \left(P_\theta(\theta, \phi) \, p_\theta(\theta, \phi) + P_\phi(\theta, \phi) \, p_\phi(\theta, \phi) \right) \sin\theta d\theta d\phi}{P_V + P_H} \qquad (5.16)$$

Two new variables dependent on θ and ϕ have also been introduced, $p_\theta(\theta, \phi)$ and $p_\phi(\theta, \phi)$. These represent the angle of arrival power density due to the incoming $E_{i\theta}$ and $E_{i\phi}$ fields at any given angle. The power densities, $p_\theta(\theta, \phi)$ and $p_\phi(\theta, \phi)$, are due to the radio environment, not the antenna and they are not necessarily equal. Therefore the distribution of $E_{i\theta}$ and $E_{i\phi}$ fields are not necessarily the same. We can consider these functions to be equivalent to a probability density function of a random variable so they will satisfy the following equations:

$$\int_{-\pi}^{\pi} \int_0^\pi p_\theta(\theta, \phi) \sin\theta d\theta d\phi = 1 \qquad (5.17)$$

$$\int_{-\pi}^{\pi} \int_0^\pi p_\phi(\theta, \phi) \sin\theta d\theta d\phi = 1 \qquad (5.18)$$

Equation (5.16) can then be modified to the following general expression for MEG:

$$\text{MEG} = \int_{-\pi}^{\pi} \int_0^\pi \left(\frac{\text{XPR}}{1 + \text{XPR}} G_\theta(\theta, \phi) p_\theta(\theta, \phi) + \frac{1}{1 + \text{XPR}} G_\phi(\theta, \phi) p_\phi(\theta, \phi) \right) \sin\theta d\theta d\phi$$
$$(5.19)$$

Observing Equation (5.19), there could be a scenarios at angles θ and ϕ, the value of G_θ is high while p_θ is low, or in vice versa. The same consideration could be made for G_ϕ and p_ϕ. If all angles of θ and ϕ are considered and there is a strong majority of cases where G_θ and p_θ, or G_ϕ and p_ϕ are high, then the MEG will also be high. If neither case occurs then the MEG will be low. Therefore MEG considers whether the antenna has high gain in the directions where the E-fields are incident onto it. The polarisation of the antenna and the incoming E-fields, however, must also be aligned. The result is then a mean gain that

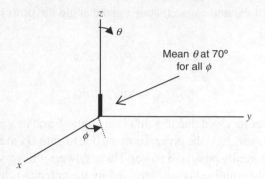

Figure 5.7   Illustration of the angle of arrival model used by Taga.

the antenna has when in an environment. Such a value is directly relevant to how much the antenna will increase or reduce the received signal at a mobile.

In order to evaluate the MEG, it is necessary to have a model for the angle of arrival functions, $p_\theta(\theta, \phi)$ and $p_\phi(\theta, \phi)$, which will depend on how scatterers are positioned. A simple example is used by (Taga 1990), which considers that there is a uniform angle of arrival in azimuth, but a Gaussian angle of arrival in elevation. We can visualise this in Figure 5.7, where the signals will arrive uniformly around the antenna but in elevation most of them will arrive above the horizontal xy plane where the mean of the Gaussian arrival is, that is, $\bar{\theta} \approx 70°$. It is therefore possible to easily define the Gaussian angle of arrival as a function of θ independent of ϕ as follows where the mean angle of arrival in elevation is defined as $\bar{\theta}$ and its standard deviation, σ_θ and σ_ϕ for the two polarisations:

$$p_\theta(\theta, \phi) = K_\theta e^{\frac{(\theta - \bar{\theta})^2}{2\sigma_\theta^2}} \tag{5.20}$$

$$p_\phi(\theta, \phi) = K_\phi e^{\frac{(\theta - \bar{\theta})^2}{2\sigma_\phi^2}} \tag{5.21}$$

To make the equations valid for calculating MEG, the constants K_θ and K_ϕ are defined to satisfy the criteria in Equations (5.17) and (5.18):

$$K_\theta = \frac{1}{\int_{-\pi}^{\pi} \int_0^{\pi} e^{\frac{(\theta - \bar{\theta})^2}{2\sigma_\theta^2}} \sin\theta d\theta d\phi} \tag{5.22}$$

$$K_\phi = \frac{1}{\int_{-\pi}^{\pi} \int_0^{\pi} e^{\frac{(\theta - \bar{\theta})^2}{2\sigma_\phi^2}} \sin\theta d\theta d\phi} \tag{5.23}$$

According to measurements by (Taga 1990), $\bar{\theta}$ would often be the same, around 70° in an urban environment. Standard deviations on the angle of arrival, σ_θ and σ_ϕ, would be between 20° and 40° in an urban environment. For purposes of ease of analysis in this chapter, we will assume that $p_\theta(\theta, \phi)$ and $p_\phi(\theta, \phi)$ are equal even though there are cases where this is not typical. The angle of arrival is not always uniformly distributed in azimuth though for simplicity the rule will be adopted in this chapter.

5.2   Single Antenna Design

It is assumed by many that if an antenna has a matched impedance (i.e. $Z_{in} = Z_0$) giving high return loss then the antenna will always radiate energy. An antenna with a good impedance match may have a low efficiency, which means the antenna will not radiate much power. If there is no efficiency there is no gain. In order to determine whether the antenna is both impedance matched and efficient, it is best illustrated by the example in Figure 5.8 used by Saunders and Aragon-Zavala (2007), where Z_{in} is split into three components: the radiation resistance, R_{rad}, the loss resistance R_{loss} and the input reactance, X_{in}. If Z_{in} is to be matched to Z_0 then ideally X_{in} should be zero to give minimum reflection coefficient in Equation 5.9. However, the resistive part of Z_{in}, even if it is the same as Z_0, will consist of two resistances added together, the radiation resistance, R_{rad} and the loss resistance, R_{loss}. If R_{rad} is negligible, then the antenna will have low efficiency because power is lost through R_{loss} where as if R_{loss} is negligible then the antenna is highly efficient so it will radiate or receive energy. In both cases, however, Z_{in} still has the same value and is still matched to Z_0 so Z_{in} alone cannot tell us how efficient the antenna is. The only way to determine the efficiency in this regard is by knowing R_{rad} and R_{loss} (which can not be directly measured) as:

$$\text{Efficiency} = \frac{R_{rad}}{R_{rad} + R_{loss}} \qquad (5.24)$$

Clearly, it can be seen that an antenna with good impedance match is not necessarily efficient since antenna impedance and efficiency are mutually exclusive. Antenna design therefore requires constructing an antenna where radiation resistance is close to the input impedance (if it is resistive), which can be difficult. The simplest example of how to design an antenna in such a way is a half wavelength dipole. Theoretically it can be shown that when it is half a wavelength it has a high radiation resistance of 73Ω and an input reactance

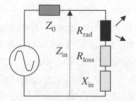

Figure 5.8   Diagram showing the radiation resistance and loss resistance of an antenna.

of 42.5Ω giving an input impedance of $73 + 42.5j\Omega$ (Balanis 2005). This impedance is not close enough to a source impedance of 50Ω, so to remove the reactive component, the dipole can be shortened at both ends by 5–10%. The input resistance then reduces to around 60Ω, which gives a suitable return loss of at least 10 dB but slightly higher loss resistance. If the dipole was shortened any further, there may be chance that the impedance is closer to the desired impedance of 50Ω but the loss resistance would increase and therefore efficiency would reduce. It is therefore the case that the 5–10% shortening of the dipole provides a compromise where the efficiency is slightly less than maximum so the input impedance is close enough to the required value.

Mobile antennas often require re-designing dipole or monopole antennas so as to fit the size requirements. Size can often be a limitation and when trying to consider both impedance and efficiency, the antennas will often be far from ideal (Chen *et al.* 2007). Furthermore, interaction with the user or surrounding immediate area (known as near field effects) can impact the impedance and gain, which must also be considered when designing antennas. Many antennas also have to operate over a suitable bandwidth. This therefore means that antennas should be tested both for their return loss and gain over the desired frequency range to determine what bandwidth they have.

Fundamentals of single antenna design

For any antenna to radiate or receive power, it will require a suitably matched impedance. If the antenna's impedance is matched it will not necessarily radiate or receive power, because its efficiency is not dependent on impedance. Therefore if the efficiency is not also maintained, the antenna will be of poor quality.

In a real environment, the antenna will be capable receiving electric fields in certain directions and polarisations that may not be consistent with the polarisations and angles of arrival of the electric fields. This will determine how high the antenna's mean effective gain (MEG) is. To determine cases where an antenna has high mean effective gain, the following principles can be applied:

▶ Where the values of, $A_\theta(\theta, \phi)$ and $E_{i\theta}(\theta, \phi)$, while also the values of $A_\phi(\theta, \phi)$ and $E_{i\phi}(\theta, \phi)$ are at a maximum for the same angles of θ and ϕ, then there will be a high MEG. If they are opposite then the MEG will be low. Therefore high MEG is determined by where an antenna will receive electric fields in the direction in which they arrive but also with the same polarisation.

▶ A vertically polarised antenna within an environment that has a low XPR will have a low MEG; however, for a high XPR, the MEG will be dependent on the antenna pattern and the angles of arrival of the incoming fields.

▶ A horizontally polarised antenna within an environment that has a high XPR will have a low MEG, however for a low XPR, the MEG will be dependent on the antenna pattern and the angles of arrival of the incoming fields.

5.3   Designing Array Antennas for MIMO

The correlation between the antenna elements of an array is of the key contributions to the performance of MIMO antennas. Therefore, before using antennas for MIMO, they need to be designed in a way that their elements will receive de-correlated signals, which will be a summation of all the incoming E-fields as well as the antenna's radiated E-field pattern. It is possible to do this in different ways such as spatially separating the antennas or changing their angular patterns or polarisations, which will be explained in the following sub-sections.

5.3.1   Spatial Correlation

In the general case, the complex correlation between two antennas m and n, can be derived using the radiated E-field patterns $A_{\theta m}$, $A_{\phi m}$, $A_{\theta n}$ and $A_{\phi n}$. The terms A_θ and A_ϕ are the same as those applied in the previous section, the only difference being that they have been assigned to the antenna number. The magnitude of these, when considered as relative values to a fixed angle, is equal to the square root of their corresponding gains, $G_{\theta m}$, $G_{\phi m}$, $G_{\theta n}$ and $G_{\phi n}$, though phase information would need to be added in to make them complex, which can be found by antenna measurement or simulation. Using these amplitudes, the complex correlation, ρ_{mn} is defined as follows (Brown $et\ al.$ 2005):

$$\rho_{mn} = \frac{\int_{-\pi}^{\pi} \int_0^{\pi} \left(\text{XPR} A_{\theta m}(\theta, \phi) A_{\theta n}^*(\theta, \phi) p_\theta(\theta, \phi) + A_{\phi m}(\theta, \phi) A_{\phi n}^*(\theta, \phi) p_\psi(\theta, \phi) \right) \sin\theta d\theta d\phi}{\sqrt{\sigma_m^2 \sigma_n^2}}$$

(5.25)

where

$$\sigma_n^2 = \int_{-\pi}^{\pi} \int_0^{\pi} \left(\text{XPR} |A_{\theta n}(\theta, \phi)|^2 p_\theta(\theta, \phi) + |A_{\phi n}(\theta, \phi)|^2 p_\phi(\theta, \phi) \right) \sin\theta d\theta d\phi \qquad (5.26)$$

and the same equation is valid for antenna m. The correlation is therefore comparing the radiated E-field patterns (which are reciprocal whether the antenna is in transmit or receive mode) with the incident E-fields arriving at the antenna, which are defined by the same probability equations, $p_\theta(\theta, \phi)$ and $p_\phi(\theta, \phi)$, used for MEG. The antennas may have different spacing, different radiated E-field patterns or different polarisations. In all cases, the radiated E-field patterns will be different as the phases between the two antennas will be different even if their magnitudes are the same. First of all we will consider two spatially separated omnidirectional antennas, where the radiated E-fields have horizontal spacing, d and vertical spacing, l. In this instance if we assume the antennas are both purely vertically polarised ideal antennas so all E-fields have the same magnitude, (Brown $et\ al.$ 2005) has derived the spatial correlation for vertical and horizontally separated antennas:

$$\rho_{mn} = \int_{-\pi}^{\pi} \int_0^{\pi} e^{j\beta \sqrt{d^2 + l^2} \cos\xi} p_\theta(\theta, \phi) \sin\theta d\theta d\phi \qquad (5.27)$$

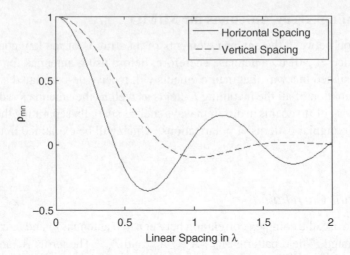

Figure 5.9 Graph showing the correlation between two adjacent antenna elements when vertically and horizontally spaced.

where

$$\cos\zeta = \sin(\theta + \delta\theta\,\mathrm{sgn}\phi)\sin\phi \qquad (5.28)$$

$$\delta\theta = \tan^{-1}\left(\frac{l}{d}\right) \qquad (5.29)$$

This equation is plotted in two ways in Figure 5.9 where just vertical spacing l (i.e. d stays zero) and horizontal spacing d (i.e. l stays zero) is considered. The angle of arrival model used by Taga as described earlier is used here, where the mean angle $\bar{\theta}$ is at $70°$ and the standard deviation σ_θ is $20°$. Due to the fact that there is a nonuniform angle of arrival power density in elevation but not in azimuth, the spacing required to get low correlation (as would be required for MIMO) is significantly greater for vertical spacing. For this reason, array antennas for MIMO are generally spaced horizontally rather than vertically so as to have more compact antenna designs.

It is also worth giving consideration as to why the antennas are spaced in a linear topology as they could for example be spaced in other topologies as follows:

▶ **Circular topology** as in Figure 5.10 where the antennas are evenly spaced in a circle with radius r_c in the horizontal plane. We can derive the antenna spacing between two adjacent elements, d, when we know the radius and angle α (derived by dividing $360°$ by the number of elements) by using Pythagoras' theorem:

$$d = r_c\sqrt{2(1 - \cos\alpha)} \qquad (5.30)$$

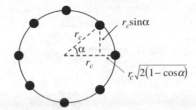

Figure 5.10   Diagram of a circular array geometry in the horizontal plane.

This can then be inserted into Equation (5.27) to calculate the correlation between adjacent elements as a function of radius r_c in Figure 5.11. It can therefore be seen that the more elements we require, the larger the radius is required to accommodate them and result in suitable de-correlation. Therefore circular arrays are not as practical as linear arrays for MIMO.

▶ **Planar topology**, which will make linear arrays even more compact using examples in Figure 5.12. All antennas have minimum spacing d in this case, which if sufficiently spaced apart (i.e. greater than half a wavelength) will allow low enough correlation between all elements.

5.3.2   Angular and Polarised Correlation

As well as using spatial separation to obtain de-correlated antenna branches, it is also possible to use angular or pattern differences and polarisation differences between the

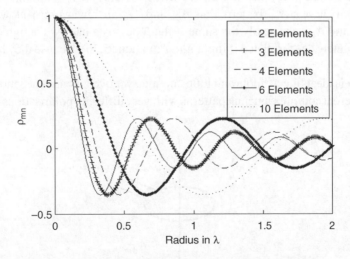

Figure 5.11   Graph comparing the spatial correlation of adjacent antenna elements for a circular array antenna with a different number of elements and varying radius.

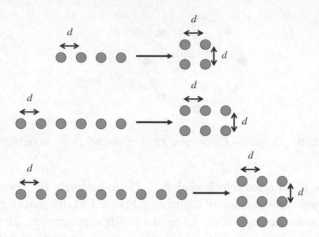

Figure 5.12   Diagram showing examples of linear arrays converted to planar arrays.

antenna elements. This is particularly important for MIMO antennas placed on small mobile terminals because they do not have enough physical size to accommodate sufficiently
spaced antennas. First of all, it must be established by inspection of Equation (5.25) that
correlation is calculated as a comparison of two complex antenna patterns in the same
polarisation. This likewise applies to two antennas that may be located close to each other
(so they have little spatial de-correlation) and have different angular patterns, that can
result in a lower angular correlation (which is also known as pattern de-correlation). Take
a dipole antenna and a loop antenna like that shown in Figure 5.13. If the antennas are
perfectly polarisation pure, then the dipole will have only a $A_{\theta n}$ pattern, while the loop
antenna will only have a $A_{\phi m}$ pattern, thus $A_{\phi n}$ and $A_{\theta m}$ are always zero. Now the complex
patterns, $A_{\theta n}$ and $A_{\phi m}$ are nearly the same if the loop has a narrow enough diameter, yet
they are completely de-correlated if input into Equation (5.26) because they have opposite
polarisation.

No antenna is ideal like a dipole and loop antenna, yet two antennas at a mobile terminal
may have different complex antenna patterns with very differing polarisations because they

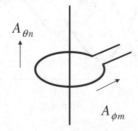

Figure 5.13   Example of a loop and dipole antenna with full polarisation de-correlation.

will have a mixture of A_θ and A_ϕ. Therefore, polarisation correlation is inherent within angular correlation (Brown, Saunders *et al.*, 2007) if one antenna receives a greater number of fields in one polarisation than the other. However, it should be noted that to exploit polarisation to get better angular de-correlation, it depends on having a suitable XPR in the channel. If the XPR is above 10 dB then there is negligible horizontal polarisation to be able to exploit. Ideally the XPR should be around 6 dB or even lower. If there is a high XPR and there is low angular correlation, it is the case because the antennas are in the same polarisation but they have different patterns.

De-correlated MIMO antennas

For two or more antennas at the Tx or Rx, spatial separation is one of the simplest ways to gain suitable de-correlation. If the antennas are at least half a wavelength apart then theoretically they will be suitably de-correlated to support a MIMO system. It is possible that the antennas could be spaced apart in a long line, though it is also possible they can be arranged in several rows or columns, which results in a planar antenna array.

At the mobile terminal, the spatial separations required are often too large to be accommodated. Therefore it is necessary to find other means of creating to antennas that have de-correlated outputs. One such example is to use different polarisations. The ideal case where there is complete de-correlation between the two polarisations is to use a vertically polarised antenna (which could be a dipole) and a horizontally polarised antenna (which could be a loop). In this case, there would be complete polarisation de-correlation, though also there would be angular de-correlation.

The angular patterns of two different antennas may have different patterns that result in a de-correlation, which may also be due to their differences in polarisation. Therefore, the angular correlation is defined as the difference between two antenna patterns in the same polarisation. It may be the case that the antenna patterns are very similar (such as a dipole and loop), but their polarisations are opposite, therefore their angular correlation is low.

5.3.3   Impact of Nonuniform Angles of Arrival

The results presented for correlation in the previous sections have been computed using a uniform azimuth and Gaussian elevation angle of arrival at the mobile, which is a simple enough model to obtain results mathematically. It is not necessarily the case that a uniform azimuth angle of arrival is a suitable model for the radio environment at a mobile Rx (COST259 2001). For example, the location of the mobile may be within a corridor in the indoor case or within a street canyon in the outdoor case. Though both instances are

NLOS, the angle of arrival is not uniform since the most scatterers are at the sides of the street/corridor, not along it. Therefore there are two main angles of arrival from the two sides opposite each other.

In mobile radio environments generally, the elevation angle of arrival is also considered sometimes to be a different distribution such as Laplacian rather than Gaussian (Ertel and Reed 1999) and highly dependent on the height of the Tx antenna. If there is a line of sight, this will inevitably create a point from which there is a single constant angle of arrival, with some minor arrival angles near to it. The following factors will have a strong influence on the AOA distribution in a given environment:

▶ **The polarisation of the Tx source:** Several measurements have verified that if the Tx source is vertically polarised, the angle of arrival is significantly different to when it is horizontally polarised. This is because incident waves on a scatterer will de-polarise differently depending on what polarisation they had incident on the scatterer. Furthermore there will be different resultant angles of arrival if both vertical and horizontal polarisations are transmitted at the same time.
▶ **Frequency dependence:** Angle of arrival is highly frequency dependent because the scatterers will reflect signals differently over frequency. Therefore it is necessary to have information about statistics of different radio environments for different frequency bands of operation.
▶ **The respective heights and distances between the antennas:** The antenna height can be particularly important, especially at the Tx and also the distance between the Tx and Rx. Several measurements and models have been made to derive suitable models for different applications

Correlation and MEG are not always evaluated just by mathematical models of AOA, it is also the case that real data can be directly input to evaluate such information about array antennas. In the case of Nonuniform AOA in azimuth, it means that correlation and MEG are dependent on the orientation of the Rx and Tx, which could be highly variable. Therefore it is not possible to simply take one value of correlation or MEG in such cases, more often a range needs to be analysed and then averaged. As the MEG and correlation are evaluated in discrete form, they require the antenna patterns and angle of arrival models to have a suitable resolution. Research has shown that for mobile terminal antennas, antenna pattern cuts should be taken in 10° steps, which is sufficient to give accurate results (Sulonen *et al.* 2002). However if the antennas had high directivity, a higher resolution would be required though this does not normally occur at the mobile.

5.4 Impact of Antenna Design on the MIMO Radio Channel

Up to this point, it has been assumed that the antennas have no mutual coupling effects. When antennas are closely spaced together, some current (and likewise power) will cross couple from one antenna element into neighbouring antenna elements. The closer the

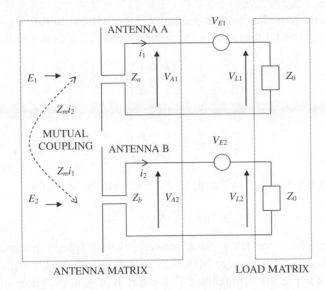

Figure 5.14 Diagram showing the principles of mutual coupling between two antennas.

antenna elements are, the more vulnerable they are to this. For multiple elements, the coupling can be determined in terms of an N-port Z-parameter impedance matrix (Brown, Saunders *et al.* 2007; Vaughan and Andersen 2002), where N is the number of antenna ports, to represent the impedances between the antennas that are represented by complex impedance matrices **Z**. The impedance of each antenna element is a combination of two types of impedance, self and mutual impedance. The self impedance is the ratio of voltage to current in the antenna at the antenna input port. The mutual impedance is the ratio of the voltage at the antenna to the current at each of the other antennas, when the antenna at the input port is excited. Therefore, each input port will have one self impedance and $N - 1$ mutual impedances, giving $N \times N$ impedances. These impedances for two antennas are illustrated in Figure 5.14, where there are E-fields incident on the two antennas, A and B, given as E_1 and E_2. These will then induce current on both antennas, i_1 and i_2, which create voltages, V_{E1} and V_{E2}. The antennas therefore have two self impedances, Z_{inA} and Z_{inB} and two mutual impedances that have the same value Z_m. Therefore the complete antenna impedance matrix in the 2x2 case, $\mathbf{Z}_{ANT}$ and the load impedance, $\mathbf{Z}_L$, which represents the load applied to the two antennas, are defined as follows:

$$\mathbf{Z}_{ANT} = \begin{pmatrix} Z_{inA} + Z_m & Z_m \\ Z_m & Z_{inB} + Z_m \end{pmatrix} \tag{5.31}$$

$$\mathbf{Z}_L = \begin{pmatrix} Z_0 & 0 \\ 0 & Z_0 \end{pmatrix} \tag{5.32}$$

Dependent on the impedances in $\mathbf{Z}_{ANT}$ and $\mathbf{Z}_L$, this will cause varying voltages V_{ANT1}, V_{ANT2}, V_{L1} and V_{L2} so that they can be incorporated into vectors as follows:

$$\mathbf{v}_E = \begin{pmatrix} V_{E1} \\ V_{E2} \end{pmatrix} \tag{5.33}$$

$$\mathbf{v}_L = \begin{pmatrix} V_{L1} \\ V_{L2} \end{pmatrix} \tag{5.34}$$

$\mathbf{v}_E$ and $\mathbf{v}_L$ are related as follows (Vaughan and Bach Andersen 2002):

$$\mathbf{v}_L = \mathbf{v}_E \mathbf{Z}_{ANT}(\mathbf{Z}_{ANT} + \mathbf{Z}_L)^{-1} \tag{5.35}$$

The mutual coupling that takes place between closely spaced antennas can be both an advantage and a disadvantage to MIMO. The mutual coupling if high will cause the antenna elements in the array antenna to have different patterns compared to what the patterns would be if the elements had no mutual coupling. The reason for this is that when a current is flowing into one antenna, some of that current will also flow onto a nearby antenna due to the mutual impedance. Such antenna is known as a parasitic element. As has been realised from the beginning of this chapter, the current in the conducting element will form an E-field and in this case, both the antenna connected to the source and the parasitic antenna will radiate a resultant E-field together. Therefore, the radiated pattern would be based on a superposition of the E-field generated by these two antennas. What would then happen is that when one antenna is excited, a pattern in one direction would result, while when the other antenna is excited, a different pattern in another direction would result. This would therefore mean that the two antennas would have a certain angular correlation and if the mutual coupling caused the antennas to have different enough patterns, this correlation would be low. While mutual coupling may help reduce correlation, at the same time there is less current in the driven antenna and the radiation resistance reduces, which reduces the efficiency in the antennas. The impact of the mobile terminal array antenna, both due to its inherent mutual coupling and interaction with the user, will impact the resultant channel state arriving at the receiver, in a way that the maximum channel capacity possible will be limited due to the antenna. The impact on the wireless channel due to the antenna is summarised as follows:

▶ **Degradation of antenna efficiency and MEG:** If two monopoles are brought close together, the mutual coupling will cause their efficiency to degrade which will likewise degrade their MEG both due to the loss but also because they are less omnidirectional (Brown 2004). This loss will have a direct degradation on received SNR, which will likewise offset the capacity to a lower level.

▶ **Antenna mismatch:** As shown in Equation (5.31), the mutual coupling is added to the self impedance of the antenna. Therefore, when one antenna is in the presence

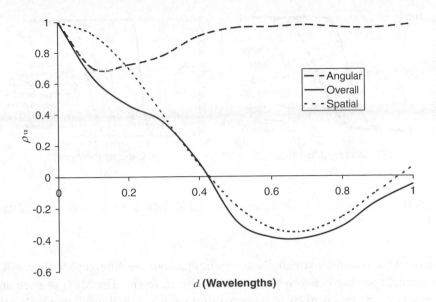

Figure 5.15 Comparison of the angular and spatial correlation of two spatially separated monopoles based on the Taga angle of arrival model; Source: Brown, Saunders *et al.* (2007). Reproduced by permission of © 2007 IET.

of another it will have a different impedance, which is possibly not comparable of that of the source impedance. It is often therefore necessary to adjust the antennas to improve their impedance match when they are in the presence of other antennas, though as explained earlier, this can be at the expense of degrading the efficiency of the antenna. This may require compromise if the MIMO array antenna has restrictive size constraints.

▶ **Correlation between two antennas:** It can be seen in Figure 5.15 that for two monopoles (which have the same characteristics as half wavelength dipoles), spacing of less than 0.5 wavelengths allows angular de-correlation to predominate where as the spatial de-correlation dominates beyond 0.5 wavelengths. Therefore, the correlation overall is below 0.7 from only 0.1 wavelengths distance due to the mutual coupling. This is desirable for MIMO so therefore effects of mutual coupling are beneficial in this case. It is assumed in Figure 5.15 that the angle of arrival model is that used by Taga (1990) explained earlier in this chapter.

▶ **Branch Power Ratio (BPR):** The mutual coupling may cause two antennas to radiate in different directions but at the same time (depending on the angle of arrival) the MEG of the two antennas may be considerably different. This will indicate that the branch power ratio (BPR) of the two antennas will be high, which is defined as the ratio of power in one antenna to the other. Antennas at the mobile in particular are subject to user interaction, for example when the user is holding the mobile terminal to have a telephone conversation or holding the handset in front of them to send a text message or

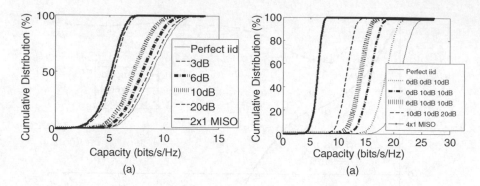

Figure 5.16   Illustration of Rx BPR impact on iid MIMO channels for the (a) 2x2 case and (b) 4x4 case.

observe what is on their screen. Such user interactions are different and they will cause the mutual coupling between the antennas to be different. The MEG of each antenna branch in such cases is likely to be very different, which will result in a high BPR.

Figure 5.16 (a) illustrates a 2x2 MIMO case, where the BPR at the mobile of two independent Rayleigh channels will shift the capacity down to that of a 2x1 MISO channel if the BPR goes beyond 10 dB. BPR beyond 6 dB has significant degradation in capacity. An example of the same scenario with a 4x4 case is shown in Figure 5.16 (b), where more branches help to reduce the impact on capacity if one of the branches is weak. In this case there are three BPRs: from branch 1 to branch 2, from branch 1 to branch 3 and branch 1 to branch 4. If more than two branches are weak then capacity begins to drop significantly down to that of a 4x1 MISO channel though less than half the capacity is lost when only two branches are weak.

From analysing the above four points, compact MIMO antennas may benefit from mutual coupling by being able to decrease the correlation between branches, but at the same time they are vulnerable to having high BPRs and degradation in efficiency and MEG which will counteract the benefits. A limit on mutual coupling is therefore required which will ensure the conflicting factors' balance.

Compact antennas and mutual coupling

Mutual coupling between closely spaced antennas can improve de-correlation to reduce it to be less than the theoretical spatial correlation. However, if the spacing is too close, the antenna's efficiency can be compromised, which will therefore reduce the received SNR. This reduction in SNR will outweigh the benefit that a low correlation between the antennas would bring to the MIMO system. Therefore a minimum spacing

Compact antennas and mutual coupling (Continued)

would have to be maintained so that the benefits of mutual coupling between the antennas could still be gained.

At the mobile, the impact of the user on mutual coupling when using the terminal must also be considered as well as the fact that the body will absorb significant amounts of radiated power if the antenna is designed badly. One common method to mitigate these problems is to use antennas that will typically radiate away from the user when in use. However, this can be difficult if the mobile terminal has more than one mode of usage (e.g. as a mobile handset or a handheld multimedia device).

User impact will also have a bearing on the directions in which the mobile terminal antennas can radiate and receive electric fields as well as their polarisation, which will impact the MEG. The MEG may also not be consistent from one antenna to the next, thus there will be a resulting branch power ratio, which should be as close to 0 dB as possible.

5.5 Evaluating Antenna Impact on the MIMO Channel

This section first shows, by using a crude example, how antennas can impact capacity by using just correlation, MEG and BPRs etc. from the antennas. This does not provide accurate evaluation, however, of the antenna's MIMO potential. The second part of this section therefore provides an explanation of some advanced techniques to carry out analysis of MIMO array antennas.

5.5.1 A Crude Evaluation of the Impact of Antennas on MIMO Channel Capacity

It can be shown from (Loyka and Tsoulos 2002) that if the Tx is assumed to have de-correlated branches, the correlation at the Rx can be used directly to give a crude estimate on the maximum achievable capacity, C, where it is assumed that there is no BPR in this case and the radio environment is independently identically distributed before antenna effects are applied as follows:

$$C = \log_2 |I + \text{SNR}\,R_{\text{Rx}}| \tag{5.36}$$

where SNR is defined as the signal to noise ratio and I is the identity matrix. In the case of 2x2 MIMO, the correlation matrix is as follows:

$$R_{\text{Rx}} = \begin{pmatrix} 1 & \rho_{12} \\ \rho_{12}^* & 1 \end{pmatrix} \tag{5.37}$$

so therefore Equation (5.36) resolves to be:

$$C = \log_2 \begin{vmatrix} 1 + \text{SNR} & \text{SNR}\rho_{12} \\ \text{SNR}\rho_{12}^* & 1 + \text{SNR} \end{vmatrix} = \log_2 \left[(1 + \text{SNR})^2 - \text{SNR}^2 |\rho_{12}|^2 \right] \qquad (5.38)$$

This equation can also be modified to include a loss factor, L_f, which is the ratio of the MEG of any monopole when it is in the presence of the other to the MEG when the other is not present. When the monopoles are close together, the MEG degrades and the loss factor is as high as 10 dB so this needs to be incorporated as follows:

$$C = \log_2 \left[(1 + \text{SNR}.L_f)^2 - \text{SNR}^2 L_f^2 |\rho_{12}|^2 \right] \qquad (5.39)$$

Taking the correlation values as plotted in Figure 5.15 as well as the and the degradation values of L_f versus distance taken from Brown, Saunders *et al.* (2007) the resultant capacity versus distance can be derived and its values are plotted in Figure 5.17. Though there is suitably low correlation as low as 0.1 wavelengths apart in Figure 5.15, the value of L_f is below −2 dB so it results in the capacity being only 50% more rather than near to 100% more than the SISO capacity. With the spacing at least 0.5 wavelengths, the value of L_f increases towards 0 dB and the correlation is negligible so the capacity is roughly constant and almost twice the SISO capacity as would be expected in the ideal case.

This sub-section illustrates how closely spaced compact MIMO antennas with mutual coupling can both be a friend or foe to MIMO systems. Correlation may be reduced, but at cost of reducing antenna efficiency. Other effects such as the BPR, which can result

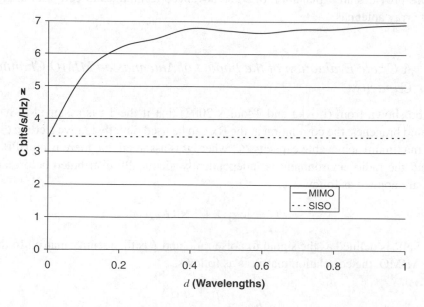

Figure 5.17   Illustration of the capacity limits for two horizontally separated monopoles.

from user interaction with mobile antennas will cause further degradation in capacity. It is therefore not sufficient to use antenna correlation as the only figure of merit to evaluate the antenna's potential for MIMO. Other methods are therefore necessary to analyse the inter-dependence between branches and also in cases where the channels are not separable, the antenna designs at both link ends will matter to the overall output as had been seen in the previous chapter. More detailed methods are therefore needed to determine how different antenna designs impact the MIMO link.

5.5.2 Advanced Techniques to Evaluate MIMO Antenna Performance

To fully accommodate all these effects, it is possible to incorporate data from the antenna field patterns (when the antennas are simulated/measured in the presence of each other) into a ray tracing model or dual directional channel model. However, as was seen in the previous chapter from the channel models themselves, the scale of data required and length of computation is high and incorporation of antenna patterns will increase this several times. Furthermore, these results would only evaluate the antenna performance in a single channel scenario so several channel scenarios would be required in order to obtain sufficient statistics to calculate an average performance.

There is therefore a need to have methods of evaluation that do not require a high computational load and allow the antenna designer to make computationally efficient evaluations of antennas in order to compare the MIMO potential of different prototypes. A method recently developed to conduct such an evaluation is the experimental plane-wave based method (EPWBM) (Suvikunnas *et al.* 2006). This method takes the incoming E-fields and the complex antenna patterns as we have used in this chapter and then makes them into vectors. For simplicity we will consider a 2x2 MIMO system, where in this case, there will be radiating complex antenna patterns represented at the ath discrete sampled angle. The vectors $\mathbf{a}_\theta(a)$ and $\mathbf{a}_\phi(a)$ are therefore used to represent the radiated E-field patterns in both polarisations of two elements at the Rx below as follows:

$$\mathbf{a}_\theta(a) = \begin{pmatrix} A_{\theta1}(\theta_a, \phi_a) \\ A_{\theta2}(\theta_a, \phi_a) \end{pmatrix} \quad \mathbf{a}_\phi(a) = \begin{pmatrix} A_{\phi1}(\theta_a, \phi_a) \\ A_{\phi2}(\theta_a, \phi_a) \end{pmatrix} \tag{5.40}$$

The incoming E-fields can consequently be represented in the same way as before but also incorporated into vectors, $\mathbf{e}_{i\theta}$ and $\mathbf{e}_{i\phi}$, though this time we must consider the fact they change as the mobile moves so they are not only a function of the ath discrete angle but the nth discrete sample in time. The incoming E-fields are therefore shown as follows and they are illustrated in Figure 5.18 for clarity.

$$\mathbf{e}_{i\theta}(a, n) = \begin{pmatrix} E_{i\theta1}(\theta_a, \phi_a, n) \\ E_{i\theta2}(\theta_a, \phi_a, n) \end{pmatrix} \quad \mathbf{e}_{i\phi}(a, n) = \begin{pmatrix} E_{i\phi1}(\theta_a, \phi_a, n) \\ E_{i\phi2}(\theta_a, \phi_a, n) \end{pmatrix} \tag{5.41}$$

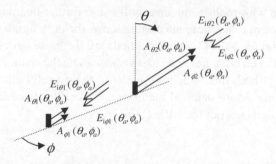

Figure 5.18   Illustration of the incident E-fields and antenna E-fields for a 2x2 MIMO case.

Therefore, the E-fields from the environment, $\mathbf{e}_\theta$ and $\mathbf{e}_\phi$ will then superimpose onto the complex antenna patterns, $\mathbf{a}_\theta$ and $\mathbf{a}_\phi$ and they will create a resultant 2x2 MIMO channel for every n^{th} time/distance variant sample. When all the resultant superimposed fields for every m^{th} polar sample in the equation below are summed together they create $H(n)$ which is time/distance variant for every n^{th} sample. This can be easily achieved using straight forward matrix algebra and N_{3DA} angular samples.

$$H(n) = \sum_{a=1}^{N_{3DA}} \mathbf{e}_{i\theta}(a,n)\mathbf{a}_\theta^H(a) + \mathbf{e}_{i\phi}(a,n)\mathbf{a}_\phi^H(a) \tag{5.42}$$

This channel output can then be used to calculate channel capacity and perform eigen analysis to determine the MIMO potential of the antenna. The test could be repeated on different antennas to make comparisons of MIMO capacity of the antennas that is repeatable. The EPWBM method still requires significant chunks of data in $\mathbf{e}_{i\theta}(a,n)$ and $\mathbf{e}_{i\phi}(a,n)$ though it is significantly less than using ray tracing or dual directional channels. The method can also be used for MIMO systems with more antenna elements though of course more antenna pattern data and channel matrices would be involved.

Impact of the antenna design on the MIMO channel and its capacity

The correlation, impedance, MEG and BPR are all useful factors to indicate the likelihood of the MIMO antenna's reliability and gain some initial indications at design stage. In the ideal case, the MEG needs to be high, the BPR close to 0 dB, the antenna impedances need to be matched to the source or receiver and the correlation between the antennas needs to be low. If any one of these criteria is not met, the MIMO capacity will be degraded due to the antennas.

To make comparable and repeatable tests between antenna designs and prototypes as well as gain some more reliable idea of their MIMO potential requires an evaluation such as the EPWBM method. Using this principle will fully accommodate

Impact of the antenna design on the MIMO channel and its capacity (Continued)

the interdependence between the MIMO paths while also being able to determine the impacts of the antenna design when superimposed onto the incoming electric fields of a specific radio environment. This is highly useful in antenna design to determine how a particular antenna is more advantageous over the other.

5.6   Challenges in Compact MIMO Antenna Design and Examples

The previous sections have identified ways in which MIMO channel capacity can be impacted due to the design of the antenna as well as ways in which the level of impact can be evaluated. These impacts are important to consider with regards to channel modelling since in an ideal case it is assumed that the antennas are spaced apart either in linear or planar form at least half a wavelength in the horizontal or azimuth plane at the frequency of operation. Were this to be true then there would be a BPR of 0 dB, low correlation between each antenna element and negligible mutual coupling. The antennas in this instance can be considered near to ideal though at frequencies of 5 GHz or less (which are or will be used for MIMO) then the spacing required to be half a wavelength apart is not practical for a small mobile terminal or access point. Were four antennas crammed onto a mobile terminal, they would more likely be as close as only one tenth of a wavelength, which has shown to degrade MIMO capacity. At a mobile terminal, it is also going to be subject to interaction with the user, which will cause the BPR, antenna efficiency and also correlation to change as the user is moving or changing their handling of the mobile. In light of these limitations, the different angular patterns and polarisations of the antennas discussed earlier have to be considered as opposed to spatial separation of the antennas. The user interaction will also impact these angular patterns though if the antennas are designed appropriately, the user can be of benefit to the antenna by helping to produce a good angular de-correlation between the antennas and not degrade their efficiency. Examples of such antennas published in the literature are the dielectric loaded folded loop and a modified planar inverted-F (PIFA) antenna in Figure 5.19. In both cases these antennas are placed on a point in the mobile terminal that is generally not covered by the hand or placed next to the head. This is true in two main scenarios, in *talk position* when the user is holding the terminal next to their head to have a telephone conversation but also in *texting position* as though the user was sending a text message or interacting with information placed on the screen of their mobile. The different position or structure of each antenna element will cause different angular patterns in the presence of the user and they will also maintain an acceptable BPR and efficiency so that MIMO capacity is not degraded.

Another way of compacting the antennas is to use spiral designs such as trapeziods, spiral antennas and an intelligent quadrifilar helix antenna (IQHA) as shown in Figures 5.20 and 5.21. Such antennas can be considered to have co-located array elements though they have a radial shift from each other allowing them to be intertwined. The mutual

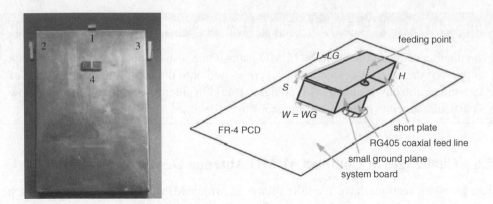

Figure 5.19 Illustration of printed and dual element PIFA antennas. Sources: Chiau *et al.* (2005) Reproduced with permission from John Wiley & Sons, Inc. and Gao *et al.* (2007) Reproduced by permission of © 2007 IET.

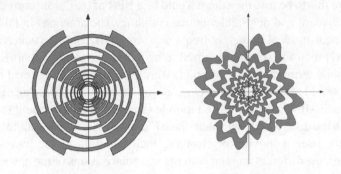

Figure 5.20 Illustration of a trapezoid spiral compact MIMO antennas; Source: Klemp and Eul (2005).

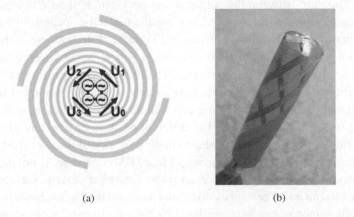

        (a)   (b)

Figure 5.21 (a) Illustration of a printed helix antenna; Source: Waldschmidt and Wiesbeck (2004); (b) an intelligent quadrifilar helix antenna. Reproduced by permission of © IEEE 2004.

coupling behaves in such a way that when one element is radiating, the others act as parasitic elements to help create a unique antenna for that element. Thus the antennas have a good angular and also polarisation de-correlation that allow the antennas to be a compact option at either the access point or mobile terminal. Since mutual coupling cannot be completely avoided in the design of such antennas as these, it is the case that they will have a capacity that is slightly lower than that of an ideal case such as the dipoles spaced half a wavelength apart though given their size is reduced to something realistic, this is a reasonable compromise.

5.7 Summary

5.7.1 Antenna Fundamentals

This chapter has provided an introduction to antenna design particularly in relation to MIMO communications. We had first established that each antenna element will need to have good *impedance matching* such that power will go into the antenna and not reflect back to the source. Once the power is in the antenna, it will need to radiate power rather than just convert it to heat or other forms of energy, thus the antenna also needs to be *efficient* which is a mutually exclusive issue from the impedance matching. A number of fundamentals about antennas were also covered in this chapter, where in particular it is important that the antennas have a *reciprocity* such that they have the same gain, efficiency and impedance matching whether they are in transmit (Tx) mode or receive (Rx) mode.

5.7.2 Designing Antenna Arrays

A MIMO antenna is required to consist of an array of several antenna elements. It was found that this is achieved best by separating the elements in a linear or planar form at least half a wavelength in the horizontal or azimuth plane. In such circumstances these antennas would be suitably de-correlated and also have a branch power ratio (BPR) of 0 dB. The mutual coupling would be negligible enough that it would not cause the efficiency of the antennas to degrade. In this case, the impact of the antennas on MIMO capacity would be negligible and they would not need to be taken into account in a channel model.

5.7.3 Practical Antennas for MIMO

When designing compact MIMO antennas suitable for a mobile terminal or small access point, the close proximity of the antenna elements will create a mutual coupling between them, where the antennas cannot be de-correlated due to their spatial separation if less than half a wavelength but the difference in their antenna angular patterns and polarisation. In such circumstances, some mutual coupling will also be present, which will slightly degrade the capacity though this cannot be completely avoided and as such is a compromise for reducing the antenna size to being physically practical for the application. The user interacting with the mobile will cause the antenna to have non ideal branch power ratio (BPR) and degradation in efficiency. Antenna design therefore is important to consider so that these impacts are minimised and examples of such antenna designs have been shown.

The following key points that the reader should have learned with regards to antenna design for MIMO are:

► Each antenna array element should have both impedance matching and efficiency though this should be considered when each element is in the presence of other array elements and the user where applicable since both these factors will impact the antenna element's own impedance due to a mutual impedance between the antennas.
► Array antennas are commonly designed as linear and planar arrays in the ideal case though for more practical and compact antenna options, the angular pattern and polarisation differences between antenna elements are used to enable the required de-correlation.
► For compact antennas the mutual impedance can be beneficial in terms of helping to create de-correlated antennas due to their angular patterns and polarisation but at the same time it causes the efficiency (and likewise the mean effective gain (MEG)) to degrade, which is a disadvantage.
► For any antenna to have potential to be used for MIMO, it is essential that there is good de-correlation between the antenna branches, low branch power ratio (BPR) whereby the MEG of the antennas should be comparable and that their efficiency is not degraded by being in the presence of each other and a user where applicable. These three factors are essential to maintain in the early stages of designing a MIMO antenna.
► For testing antenna prototypes for their MIMO potential, the experimental plane-wave based method (EPWBM) was explained as means to carry out repeatable and low complex evaluation of MIMO antennas. This works by modelling the incident E-fields onto the antenna and then considering how they will be impacted by the antenna and giving a resulting capacity.

6

MIMO in Current and Future Standards

Persefoni Kyritsi and Tim Brown

The purpose of channel models and the need to consider the effects of antennas, particularly at the mobile terminal have been outlined in the preceding chapters, which are important in testing MIMO communication systems to deliver the required capacity. As MIMO is already being used in current and future standards for mobile communications including third generation (3G) mobile, it is necessary to set standards for telecommunications equipment, both at the base station and within the mobile terminal. This will mean that when equipment is tested, appropriate channel models can be applied to ensure that they will conform to minimum standards, which mobile operators can use to assure customers that they will be delivered the service they are promised regardless of what mobile handset or device they choose. Common standards are therefore important in this regard because the handset manufacturer is normally a separate entity from the mobile operator. This chapter will outline how channel models are being used in current mobile standards and the end of the chapter will summarise applications other than mobile communications where MIMO is beginning to be exploited and as such future standards will need to be set.

6.1 Wireless Channel Modelling in Standards

As explained in Chapter 1, channel models are an abstraction of the physical world used for the computer or hardware-based simulation of physical phenomena. As such they need to achieve a balance between simplicity and fidelity, and to enable the algorithm developers, the device and equipment manufacturers to evaluate their techniques using a common platform and thus ensure objective evaluation and assessment.

Practical Guide to the MIMO Radio Channel with MATLAB® Examples, First Edition.
Tim Brown, Elisabeth De Carvalho and Persefoni Kyritsi.
© 2012 John Wiley & Sons, Ltd. Published 2012 by John Wiley & Sons, Ltd.

The channel models that were discussed in Chapter 4 were mostly link-level models, that is, they concentrated on the link between one transmitting side and one receiving side. This constitutes the first step in modelling. The next step is provided by what is known as system-level models. System level simulation is a tool widely used to understand and assess the overall system performance, taking into account the fact that the wireless system as a whole comprises several transmitting sites (e.g. base stations) dispersed over a wide area, where several receivers (e.g. mobile terminals) require access to wireless services (e.g. voice services, data services etc.). The most important considerations for system simulations are the simulation run-time, the consistency with link-level models and the statistical convergence of the results.

In a typical system level simulation, the geometry of a wireless deployment is the first to be defined (e.g. density of base stations). Then, the long-term fading behaviours and large scale parameters are derived, and then the short-term time-variant spatial fading channels are generated.

Clearly the terminals encounter different channel conditions depending on where they are: for example closer to the cell centres they might encounter stronger signal reception, at the cell edge they might be able to connect to several sites simultaneously. Therefore, system-level modelling requires that all possible link conditions along with their occurrence probability be modelled. Additionally, system models include the large scale location-dependent propagation parameters such as path loss and shadowing. Given that terminals closer together should encounter similar channel conditions with respect to path loss and shadowing, a system level model should be able to capture the relationship among links to several mobile terminals. Following a similar train of thought, since a terminal might be communicating with more than one base station simultaneously, models should be able to capture the co-dependence of links from one terminal to several base station sites.

Once the large scale parameters have been defined, there are in general two types of methodologies in the flow diagram illustrated in Figure 6.1 to generate short-term fading channels, which both fall under the stochastic channel models introduced in Chapter 4.

▶ **Ray-based modelling** The channel is constructed from summing over multiple rays which are defined geometrically. Note that this is not a ray tracing model as described in Chapter 4. In this case, the incoming rays are modelled at the receiver as a wideband impulse response with each delay tap having a given amplitude, phase time delay and angle of arrival or departure. Therefore a model is used to directly generate the impulse response statistically, which is based on measurement data. This makes the model ray-based rather than ray tracing, which will simulate the geometry to determine the incoming rays at the receiver and then the resulting impulse response. Once the impulse response data is generated for a single time instant, the amplitude and phase of the impulse response for each Tx and Rx branch combination is then adjusted based on the position of the array elements at both ends. This is very easy to do using simple trigonometry and the impulse response for each Tx to Rx branch which will change incrementally based on the incremental change in angle of arrival and departure.

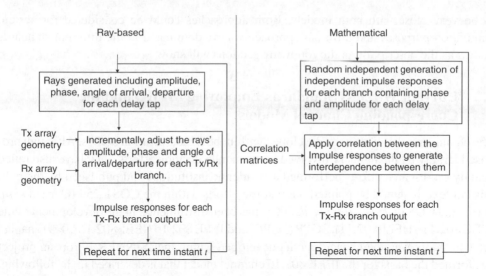

Figure 6.1   Illustration of the ray based and mathematical modelling approaches.

A receiver will perceive the channel differently depending on its array size and lay-out, its direction and speed of travel so these factors must also be taken into account. Once this has been carried out for a single time step, it must be repeated for every time step as shown by the loop illustrated in Figure 6.1. The 3GPP SCM model is such a system simulation model, as is explained later in this chapter. The ray-based models are powerful as they are independent of any particular assumption regarding the antenna configuration and can then be adapted to any defined linear or planar array configuration.

▶ **Mathematical modelling:** The other modelling methodology is mathematical or an-alytical modelling also illustrated in Figure 6.1 in which the space–time channel (i.e. impulse response as a function of both delay and spatial position) as seen by the receiver is constructed mathematically, assuming certain system and antenna parameters. In this method each channel coefficient or impulse response for every Tx and Rx branch com-bination is generated independently, but then correlated in both space and time delay in a similar way to that of the Kronecker method described in Chapter 4 so as to model their interdependence. Again, the process is repeated for each time step in a loop and the correlation values used are taken from real-world measurements. Though this model is less powerful and will inevitably compromise accuracy, its complexity is reduced.

Mathematical modelling tries to analytically model the statistical behaviour of a channel, represented by probability distributions and power profiles of delays and angles. On the other hand, in a ray-based modelling, the statistical behaviour is satisfied through the summation of multiple rays with random parameters. The two approaches can be viewed as two different simulation implementations, especially if they are based on the same probability distributions and power profiles. The system performance results are expected to be very close with both models. Both approaches could be considered for system simulation purposes and different approaches have dominated in the different standards bodies as the description in the following sections will show.

6.2   Current Wireless Standards Employing MIMO and the Corresponding Channel Models

The channel models have been developed within the standardisation bodies. Although proposed largely by the industry involved in the standardisation efforts, they have also relied heavily on research efforts performed at academic institutions and published by them. In this context, it should be pointed out that the efforts within the COST 259 (Correia *et al.* 2001) and COST 273 (Correia *et al.* 2006) has been instrumental in the development of the IEEE 802.11n (IEEE802.11), 3GPP (3GPP) and IEEE 802.16 (IEEE802.16 2009) channel models. Moreover, the channel model developed within the WINNER European project has formed the basis for the IEEE 802.16 channel model that is described in the following.

6.2.1   IEEE 802.11n

The IEEE 802.11n channel model was the first MIMO channel model to be accepted by a standards body. This was facilitated by the fact that the first MIMO products to hit the market were geared towards local area network applications, and therefore pushed the development of a MIMO channel model.

The IEEE 802.11n channel model covers a range of environments and was based on preexisting single-input single-output channel models already developed for single antenna wireless local area network scenarios, which were extended to the multiple antenna case.

The key feature was the concept of the cluster, that is, of a group of paths that have similar angles of arrival/ departure as well as similar delays. The concept of a cluster in the time domain had been introduced by Saleh and Valenzuela in (Saleh and Valenzuela 1987). Saleh-Valenzuela modelling contains channel delay taps arriving in clusters, where a cluster is defined as several delay taps with similar delay characteristics. Additionally, the Saleh-Valenzuela models specify that within each cluster the power falls off exponentially and also the power falls off exponentially from cluster to cluster. An important feature of the Saleh Valenzuela approach is that the number of clusters and the number of taps within a cluster are Poisson distributed random variables. Within the clusters the delays of the clusters (i.e. delay of the first tap of the cluster) as well as the delays of the taps within a cluster are Poisson distributed random variables. Clearly a mathematical modelling approach is being taken in this instance to generate the delay taps in the impulse responses for each MIMO branch based on a Passion distributed random process. The IEEE 802.11n channel models had as a starting point the single-input single-output (SISO) WLAN channel models proposed by Medbo *et. al.*, which covered a variety of environments (e.g. typical office environment, large open space and office environments). These measurement based, tap-delay line models involved no randomness in the number of clusters, the number of taps per cluster, the delays of the clusters and of the taps within them, as the Saleh-Valenzuela approach would suggest. Instead these were deterministically defined. However they were consistent with the cluster model proposed by Saleh and Valenzuela in the sense that they showed it was easy to even visually identify groups of channels within each channel where power falls off exponentially and also from group to group. For this reason, these models served as the starting point and distinct clusters were identified. The number of clusters vary from 2 to 6, depending on the model.

Intuitively, each cluster corresponds to a spatial area where scattering objects are concentrated, hence reflections from these objects can be expected to arrive close to one another. The contribution of the IEEE 802.11n channel model was to additionally associate each such group of scattering objects with a set of angles, that is, an angle of arrival and an angle of departure, that would conceptually identify the geometric coordinates of the group of scattering objects. Angular spread, angle-of-arrival, and angle of departure values were assigned to each tap and cluster so that they would agree with experimentally determined values reported in the literature. Specifically, the AoA and AoD of the clusters was found to be random with a uniform distribution. As an example, let us consider a situation where there are 3 clusters. The 3 mean AoAs were set by randomly generating 3 values from the 0 to 2π uniform distribution. Similarly, the 3 mean cluster AoDs were set by randomly generating 3 different values from the 0 to 2π uniform distribution, uncorrelated from the three AoA values. These 6 mean values (3 AoAs and 3 AoDs) are then fixed and used for all future channel realisations. An illustration of this in Figure 6.2 (a) where there is a wide angular spread and delay spread due to wide spacing of scatterers and also in (b) where the scatterers are close together the delay spread and angular spread are low. This illustrates how experimental results have found that as far as the cluster angular spread is concerned, it is within the 20° to 40° range and it was also observed that the cluster RMS delay spread is highly correlated (0.7 correlation coefficient) with the azimuth spread. It was also found that the cluster RMS delay spread and the azimuth spread were modelled as correlated log-normal random variables. Based on these observations, the azimuth spreads that were calculated and imposed on the models were such that models with lower RMS delay spread also have lower angular spread and models with larger RMS delay spread have larger angular spread.

Furthermore, common sense suggests that it is not necessary (and would even be unrealistically restrictive) to assume that power necessarily has stopped arriving from one group of scattering objects before power starts arriving from a new one. Instead, the signals coming from different groups of scattering objects can overlap and some taps might include more than one component, each of which corresponds to a different group of scattering objects and is coming from a different direction. In summary, although each cluster is associated with one direction, each tap can be a superposition of contributions from the current and from previous groups. The number of such components for each tap was calculated and the power of each component was also calculated, such that the total power of the tap fit the original single input–single output model, and the exponential power roll-off within each cluster was preserved.

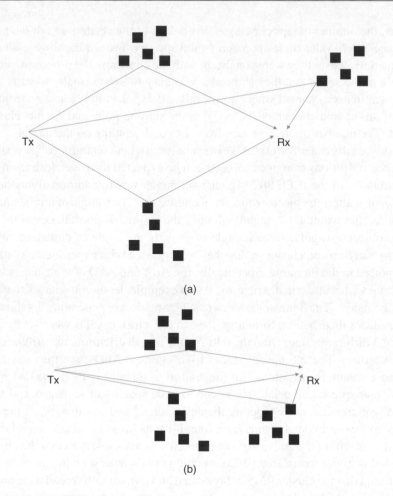

Figure 6.2 Illustration of the relation between angular spread and RMS delay spread.

In order to augment its flexibility and allow for system level simulations, the model was expanded to include:

► A line of sight component. The K factor was determined for the different environments.
► A suitable distance-dependent power roll-off law. The model that was assumed includes free space loss up to a breakpoint distance and slope of 3.5 after the breakpoint distance– for each environment a different break-point distance was determined.
► Considerations with respect to cross-polarisation. Experimentally, the XPR values were found to be in the 7–15 dB range for LOS conditions and in the 3–5 dB range for NLOS conditions. Based on these experimental results, XPR value of 10 dB was assigned to the fixed LOS matrix, and 3 dB XPR value to the variable matrix.
► A Doppler profile in order to account for time variations. The fading characteristics of the indoor wireless channels are very different from the ones we know from the

mobile case. In indoor wireless systems transmitter and receiver are stationary and people are moving in between, while in outdoor mobile systems the user terminal is often moving through an environment. As a result, a new function was defined for indoor environments in order to fit the Doppler power spectrum measurements and indeed it is centred around the zero Doppler frequency to reflect the relative temporal stability of the perceived channel.

Given the array geometry and the angular profile for each tap (number of components, power of each component, angular properties for each component), the correlations of the tap components can be analytically calculated. Indeed the IEEE 802.11n channel model is a correlation based model. In more detail, the following steps need to be performed after the calculation of the large scale parameters (path-loss, shadowing):

1. With the knowledge of each tap power, AS, and AoA (AoD), for a given antenna configuration, the correlation matrix for each tap is based on the power angular spectrum (PAS) with AS being the second moment of PAS. By using fixed values of the angles of departure and arrival, and known azimuth spectra, transmit and receive correlation matrices are computed only once.
2. We generate independent identically distributed random variables (zero-mean unit variance independent complex Gaussian random variables) and from them obtain correlated random variables using the previously calculated correlations.
3. With suitable reshaping, we obtain the channel matrix that is due to the scattered components $\mathbf{H_F}$.
4. If the tap includes a line of sight component, that is also calculated and gives rise to the channel matrix $\mathbf{H_{LOS}}$.
5. The sum of these two matrices gives the total channel transfer matrix.
6. These steps are repeated for each channel tap.

The implementation of these steps are therefore for the system developer, whose task is made lighter by the MATLAB code that was developed to accompany the standard model.

6.2.2   IEEE 802.16–WiMAX

This section focuses on the system-level simulation procedure and parameters for modelling the long- and short-term behaviour of spatial channels between a MS and one or more BSs in the context of IEEE 802.16 systems. Although such systems were originally conceived for fixed wireless applications, they have been extended to allow for user mobility and for advanced features such as relaying stations.

The general modelling approach is based on the geometry of a network layout. Therefore the parameters that are first specified relate to the network topology: cell radius, BS transmission power, BS antenna pattern, orientation, height, gain, and front-to-back ratio, MS transmission power, MS antenna pattern, height, and gain. A system level simulation typically involves the random realisation of the user placement and velocity. For system level simulations, a mix of user speeds and channel types is evaluated. Then, the long-term parameters of the link between a set of BS and a MS, such as path loss and shadowing factor, are generated according to the geometric positions of the BS and MS.

The path loss law is determined by the terrain or radio environment. For the simulation of IEEE 802.16m systems, the following scenarios are defined:

1. Urban Macrocell
2. Suburban Macrocell
3. Urban Microcell
4. Indoor Small Office
5. Outdoor to Indoor (Optional),
6. Indoor Hotspot
7. Open Rural Macrocell

The path loss models used in IEEE 802.16 channels have been based on earlier models by introducing frequency correction terms to account for the higher frequencies (3.5 GHz and above). Additionally, the path loss models employ different expressions for LOS and NLOS, while also associating a probability of occurrence for LOS vs NLOS. Moreover, the path loss models can be modified to account for situations where outdoor to indoor communication takes place and for situations as for example in a city where the cumulative distance along the city streets between the transmitter and the receiver is the distance to be taken into account instead of their absolute geometric distance.

Additionally, a shadowing is imposed such that the shadowing factor (SF) has a log-normal distribution. Its standard deviation has been derived from measurements for different scenarios. Typically, multiple links between an MS and multiple BSs are needed, among which there are multiple desired links (at least one) and multiple interference links. The shadowing factor of these links can be correlated. The site-to-site shadowing correlation is assumed to be 0.5, while the SF of closely positioned MSs is typically observed similar or correlated. Therefore, the SF can be obtained via interpolation once it has been calculated for a few discrete points, for ease of the calculation complexity.

Once the long-term parameters have been defined for the users, the short-term time-varying spatial fading channel coefficients are generated. In principle, the statistical channel behaviour is defined by some distribution functions of delay and angle and also by the power delay and angular profiles. Typically, an exponential power delay profile and Laplacian power angular profile are assumed with the function completely defined once the RMS delay spread and angular spread (both Angle of Departure (AoD) and Angle of Arrival (AoA)) are specified. The RMS delay and angular spread parameters can be random variables themselves, with a known distribution, mean and deviation. The RMS delay and angular spread can be mutually correlated, together with other large-scale parameters such as shadowing factors. Therefore there exist two approaches to the selection of the parameters:

▶ Selection according to the channels statistics

The RMS delay and angular spread parameters can be selected according to their corresponding distributions and a finite number of channel taps can be generated randomly with a per-tap delay, mean power, mean AoA and AoD, and RMS angular spread. They are defined in a way such that the overall power profile and distribution function are satisfied. Each tap is the contribution of a number of rays (plane waves) arriving at the same time (or roughly the same time), with each ray having its own amplitude, AoA, and AoD.

▶ Reduced complexity model

Clearly the above approach has higher complexity and therefore a simpler method has been selected. A reduced-complexity model is similar to the IEEE802.11n channel model described in the previous section because it specifies the delays, mean powers, and angles of the channel taps in a *pre-determined* manner. In addition to the fixed power delay, the spatial information such as per-tap mean AoA, AoD, and per-tap angular spread (thus the power angular profile) are also defined. Such models are also referred to as Cluster Delay Line or CDL models as each tap is modelled as the effect of a cluster of rays arriving at about the same time. Each tap suffers from fading in space and over time. The spatial fading process is chosen to satisfy a pre-determined power angular profile.

For each propagation scenario, the corresponding CDL model has been tabulated. It includes power delay profile (mean power of each tap and its delay) and the corresponding per-tap power angular profile (AoA and AoD).[1]

After the specification of the tap parameters, the actual realisation of a time-varying spatial channel can be performed in both ways mentioned earlier, namely using a ray-based or a mathematical correlation based approach.

Specifically for the ray-based implementation of a CDL model, each tap may be simulated via generating 20 equal-power rays, each with a known offset angles relative to the nominal tap angle of arrival or departure. The offset angles are specified in a way such

---

[1] It is useful to note that the word "cluster" is used in "clustered delay lines" in a way that deviates from its commonly accepted definition in the scientific literature, and different from the meaning it has in the context of IEEE 802.11n channels. Clusters are either defined as (i) groups of multipath components (MPCs) whose large-scale characteristics change in a similar way (e.g., as the MS moves over large distances, the relative AoAs, AoDs, and delays of the MPCs within one cluster do not change), or (ii) as groups of MPCs with similar delays, AoAs, and AoDs. For the latter definition, it is important to notice the difference between clusters and multipath groups, that is, a number of MPCs that are indistinguishable to a RX because of limited resolution are different from a cluster. A cluster consists usually of several multipath groups with similar delays and angles, and is surrounded (in the delay-angle plane) by areas of no 'significant' power. For a receiver with very low angular/delay resolution, it might happen that each cluster contains only a single multipath group, or even that a multipath group contains several clusters. Consequently, the MPCs belonging to a cluster do not change, even as the resolution of the measurement device becomes finer and finer; while the MPCs belonging to a multipath group change as the resolution becomes finer.

that by adjusting the interval between these equal-power rays a Laplacian power angular profile can be approximated. In the case when a ray of dominant power exists (i.e. when a significant Ricean component exists), the cluster has $20 + 1$ rays. This dominant ray has a zero angle offset. The departure and arrival rays are coupled randomly.

CDL models also allow for the generation of spatial correlation mathematically, which can be used directly to generate the matrix channel coefficients. The spatial correlation for each tap can be derived from the mean AoA/AoD and the Laplacian power angular profile with the specified angular spread. Per-tap correlation can also be derived numerically based on the 20 equal-power rays used to approximate the Laplacian power angular profile.

With respect to the effect of polarisation, two parameters are defined: the cross polarisation ratio XPR_V is the power ratio of vertical-to-vertical polarised component to vertical-to-horizontal polarised component, XPR_H is the power ratio of horizontal-to-horizontal polarised component to horizontal-to-vertical polarised component. It is assumed that $XPR_V = XPR_H = XPR$, where XPR is defined in Chapter 5 and the cross polarisation ratios are assumed the same for all clusters (i.e. taps). These parameters are also tabulated and given in the standard. Their values are derived from measurements.

The final parameter needed for the simulation of time-varying channels is the Doppler spectrum. If the Tx and the Rx are stationary, and the channel at time t is to be computed, then each cluster is made of a number of coherent (fixed) rays N_c and a number of scattered (variable) rays N_s (note $N_c + N_s =$ the total number of rays per cluster). The variable rays are ascribed a bell-shaped Doppler spectrum, whereas the fixed rays within a cluster share the same amplitude and phase, and their Doppler spectrum is a Dirac impulse at $f = 0$Hz.

In summary, the following procedure describes the simulation procedure based on the spatial TDL or CDL models. In the correlation based implementation, the spatial and temporal correlation needs to be derived first before generating the channel coefficients. In the ray-based approach, the time-variant matrix channels are constructed from all the rays.

▶ Step 1: Choose a propagation scenario (e.g. Urban Macro, Suburban Macro etc.). After dropping a user, determine the various distance and orientation parameters.
▶ Step 2: Calculate the bulk path loss associated with the BS to MS distance.
▶ Step 3: Determine the Shadowing Factor (SF).
   If a ray-based implementation is being used, skip steps 4, 6 and 7.
▶ Step 4: Calculate the per-tap spatial correlation matrix based on per-tap AS at the BS and at the MS, as specified in the related reduced complexity models. The spatial correlation also depends on the BS/MS antenna configurations (a random broadside direction, number and spacing of antennas, polarisation, etc.)
▶ Step 5: Determine the antenna gains of the BS and MS paths as a function of their respective AoDs and AoAs. Calculate the per-tap average power taking into account the BS/MS antenna gain.
▶ Step 6: Determine the Doppler spectrum.
▶ Step 7: Generate time-variant MIMO channels with above-defined per-tap spatial correlations.

For each tap, generate $N \times M$ i.i.d. channels first that satisfy the specified Doppler spectrum and then impose the calculated correlation values. The phases are therefore uniformly distributed.

▶ Step 8 (Ray-based method only, skip for correlation-based implementation): Generate time-variant MIMO channels as a sum of rays with the specified angular properties (known offset angles with respect to the nominal AoA and AoD). The phases are again uniformly distributed.

▶ Step 9: If a nonzero K-factor is to be enforced, the LOS path power is adjusted.

▶ Step 10: Introduce receive antenna gain imbalance or coupling, if needed.

Therefore the IEEE 802.16 channel model has several similarities with the IEEE 802.11n channel model in the sense that it is basically derived from a tap-delay line model which has been extended to include the angular properties of the taps.

6.2.3   3GPP-LTE

3GPP stands for 3rd Generation Partnership Project. The original scope of 3GPP was to standardise a 3G Mobile System based on evolved Global System for Mobile (GSM) core networks and the radio access technologies that they support (i.e. Universal Terrestrial Radio Access (UTRA) both Frequency Division Duplex (FDD) and Time Division Duplex (TDD) modes). The scope was augmented to include the maintenance and development of GSM and its evolution (e.g. General Packet Radio Service (GPRS) and Enhanced Data rates for GSM Evolution (EDGE)). Although the radio aspects of the different systems covered under 3GPP vary, a common methodology can be applied for the simulation of the physical layer and the propagation channel. Since several variants of the upcoming systems (e.g. Long Term Evolution) involve the implementation of MIMO technology, 3GPP has adopted channel models so as to specify and evaluate the algorithms and techniques proposed by the participating companies. There are two types of channel models: one that is used for link level simulations and another that is used for system level simulations.

▶ Model for link level simulations

Link level simulations reflect only one snapshot of the channel behaviour, and cannot be used to account for system-wide attributes such as scheduling applied to multiple user scenarios where each user experiences a different realisation of the channel, or for time-varying attributes such as the performance of ARQ or HARQ. However, they can be used for calibration purposes to test different implementations of the same algorithm (as opposed to implementations of different algorithms). For this reason, the model for link level simulations specifies the types of antennas used at the base station and at the mobile station, their geometric layout (linear arrays of known sizes with pre-specified element separation), as well as their orientation relative to the terminal's motion.

The model used for link level simulations in 3GPP is a tap delay line model, that is, it is assumed that the channel impulse response consists of discrete taps that arrive at known delays and to/from known angular directions with a known distribution and spread around the nominal directions. Therefore the channel can be implemented either as a sum of rays or as a correlation based model, yet satisfying in both cases the imposed statistical properties.

1. Correlation-based implementation

   As was explained in Chapter 4, the correlation of the signal received on two antenna elements can be analytically calculated from the power azimuth spectrum. Therefore this approach relies on the calculation of the correlations among the transmit/receive antenna elements based on the specified angular profiles and the generation of suitably correlated random variables. Therefore this approach resembles the correlation based modelling presented in Chapter 4.

2. Sum of rays- based implementation

   This approach is more closely related to the Jakes model (Jakes 1974), that is, each channel tap is broken into a number of rays. Each ray carries a fraction of the total tap power. For simplicity, if M rays are assumed, each one has $(1/M)^{th}$ of the total power. The directions of arrival/ departure of the rays are selected so that they satisfy the angular properties of the power azimuth spectrum.

▶ Model for system level simulations

The model for system level simulations is geared towards system-wide scenarios, and therefore implies that several base stations (cells/sectors) are transmitting to several mobile terminals. Performance metrics are collected over several realisations of the scenario, where each realisation corresponds to a simulation of a specified number of frames. For each one, the channel undergoes fast fading according to the motion of the terminals, channel state information is fed back to the base stations, then the base stations schedule their transmissions etc. Typically the number and locations are kept fixed, and what varies from one realisation to another is the initial placement of the mobile terminals. Moreover, only some of the base stations have their transmissions fully simulated including scheduling and power control, while the others are assumed to transmit at full power.

In contrast to the model for link level simulations, where the channel taps have known delays and angular properties, what is known in the system-wide model is the number of channel taps, as well as the distribution (including all the necessary parameters like mean or standard deviation where applicable), of the delay spread and the angles of arrival/departure. However the exact delays at which each path arrives and the corresponding angles for each mobile terminal are not known.

Let us illustrate this with an example.

1. Environment selection

   The environment is specified: suburban macro, urban macro, or urban micro, and the associated parameters are obtained. The environment-dependent parameters are:
   – Path loss model

- Lognormal shadowing standard deviation
- Mean total RMS Delay Spread (the delay spread is assumed to be log-normal distributed with known variance and mean)
- Delay spread (assumed to be lognormally distributed)
- Distribution for path delays (for most models this is assumed to be exponential)
- Number of paths (N)
- Number of sub-paths (M) per-path
- Azimuth spread at BS (assumed to be lognormally distributed, mean and variance provided)
- Per-path AS at BS (Fixed)
- BS per-path AoD (distribution assumed to be known)
- Mean AS at MS
- Per-path AS at MS (fixed)
- MS Per-path AoA Distribution (distribution assumed to be known)

2. Initialisation of user parameters:
   The distance and orientation parameters are calculated based on the random initial placement of the users.

3. Determine the delay spread (DS), azimuth spread (AS), and shadow fading (SF). Each one is assumed to be a log-normal distributed random variable, and they are generated so that they are correlated with known correlations. Specifically the shadow fading is also correlated from site to site.

4. Determining the parameters for the channel taps:
   - Determine random delays for each of the N multipath components. This is achieved by generating exponentially distributed random variables, such that the mean is the determined delay spread from the previous step.
   - Determine random average powers for each of the N multipath components. (Power delay profile assumed to be exponential and therefore the tap powers are determined from the delays calculated before. The randomness is introduced by imposing an additional random shadowing factor.)
   - Determine AoDs for each of the N multipath components. The AoDs are assumed to be Gaussian with a variance that depends on the azimuth spread calculated in the previous step and an additional environment-dependent factor.
   - Associate the multipath delays with AoDs.
   - Determine the AoAs for each of the multipath components. The AoAs are assumed to be Gaussian with a variance that depends on the tap power.

5. Determining the parameters for the sub-paths.
   For each path:
   - Determine the powers, phases and offset AoDs of the $M = 20$ sub-paths at the BS. The power is split equally among the sub-paths, the phases are taken to be uniformly distributed, and the offset AoDs are pre-calculated.
   - Determine the offset AoAs at the UE of the $M = 20$ sub-paths at the MS, similarly as for the offset AoDs.

– Associate the BS and MS sub-paths. These are paired randomly.
6. Relating to the geometry
   Determine the antenna gains of the BS and MS sub-paths as a function of their respective sub-path AoDs and AoAs
7. Relating small and large scale parameters
   The path loss based on the BS to MS distance from Step 2 and the log normal shadow fading determined in step 3 are applied as bulk parameters to each of the sub-path powers of the channel model.

6.2.4   Comparison of the IEEE 802.11n, WiMAX and 3GPP Models

The three models described in the previous sections have several similarities with respect to their structure and implementation. They differ in the kinds of environments they describe: the IEEE 802.11n model is geared towards indoor and small cell scenarios, whereas the other two also cover outdoor (macrocell) environments.

As explained in the previous sections, the IEEE 802.11n channel model is correlation-based, and the WiMAX and 3GPP channel models can also be implemented in the same way. The following table will be useful in highlighting the similarities and illustrating the differences of the three models (clearly all the features depend on the environment chosen):

Feature	IEEE802.11n	WiMAX	3GPP
Power delay profile	Known	Known	Exponential
Number of taps	Known	Known	Known
Delay spread	From pdp	From pdp	Log-normal
Tap delay	Known	Known	Uniform
Tap power	Known	Known	From pdp and tap delays
Angular distr. at TX	Laplacian (per cluster)	Laplacian (per tap)	Laplacian (per tap)
Angular distr. at RX	Laplacian (per cluster)	Laplacian (per tap)	Laplacian (per tap)
Mean AoD at TX	Known (per cluster)	Known (per tap)	Uniform/ Gaussian (per tap)
AS at TX	Known (per cluster)	Known (same for all taps)	Known (per tap)
Mean AoA at RX	Known (per cluster)	Known (per tap)	Uniform/ Gaussian (per tap)
AS at RX	Known (per cluster)	Known (same for all taps)	Known (per tap)

If we concentrate on the ray-based implementation of the WiMAX and 3GPP models, the following comparison arises:

Feature	WiMAX	3GPP
Number of sub-taps M	20	known
Power of each subtap	1/20	1/ M
Phase	uniform	uniform
Offset AoA or AoD	precalculated	precalculated

Summary of system level channel modelling

This chapter has outlined that channel models are set up at link-level to model the propagation scenarios that are identified through measurements, from which system-level models can be developed to determine the occurrence of both average and extreme link-level channel conditions. Using the examples given, two methods of system-level modelling have been identified:

▶ **Mathematical modelling:** In an example such as the IEEE 802.11n model, the measurement results have informed the model developers that there is a link between RMS delay spread and the AoA and AoD. This assumption has made it possible to develop correlation matrices so that the delay taps on each snapshot can be interrelated, which happens in reality as a result of the AoAs and AoDs. A further point to note is that the Doppler spread of the channel is considered which will enable the channel model to have memory when considering the relationship between one snapshot to the next.

▶ **Ray-based modelling:** This method of modelling (different to ray-tracing) is used in the 3GPP LTE model where first of all a suitable environment is selected (e.g. urban, suburban, rural). From this suitable delay spread, shadowing, path loss and angular spread (including AoAs, AoDs and their relationship) can be used as input parameters that resemble the given environment. The angles of arrival and departure for the rays can then be superimposed on to the antenna patterns that are input into the model and the respective magnitude and phase of each delay tap from each transmit to receive antenna can be determined. Therefore the rays are modified based on the antenna geometries at both ends. This is carried out for each snapshot and again like the mathematical modelling, Doppler spread can be applied so as to include memory in the channel.

The ray-based modelling method, though complex is powerful because it can be adapted to any given antenna array geometry and setup for conformance testing mobile systems.

Summary of system level channel modelling (Continued)

On the other hand, the mathematical based model makes more restricted assumptions about the antennas used though it is simpler and can more easily be implemented, which would be a suitable approach where users are not interacting with the antennas so much and so some assumptions about the antennas at both ends can be deemed reasonable.

6.3 MIMO in Other Areas

Although the standards mentioned in the previous sections already use MIMO models as MIMO transmission techniques are foreseen today for their operation, it is anticipated that in the future several other areas will require the development of corresponding MIMO channel models. So far, the inherent properties of the channel in these areas appear to be the limiting factor to the application and proliferation of MIMO systems. A bibliography has been provided at the end of this chapter where suitable references will be included for further reading.

6.3.1   MIMO for DVB-T2

DVB stands for Digital Video Broadcast and has been developed to replace conventional analog broadcasting with digital television broadcasting, and hence offer improved operational flexibility, as well as a variety of new services. Indeed, since its inauguration in 1993, digital video broadcast for terrestrial transmission (DVB-T) has responded to the objectives of its designers. However, the larger capacity requirement for high definition television (HDTV) has prompted, since the spring of 2006, the development of a future second generation of DVB-T, called DVB-T2, which is expected to include MIMO techniques. At present, DVB-T2 is geared towards plugged TV terminals or unplugged terminals in an indoor or low speed environment. It is expected that in the future, DVB-T2 and its evolution will also be geared towards highly efficient performance in environments with high Doppler spread, that is to say terminals (unplugged or not) in high speed vehicles.

DVB-T2 operates on the basis of OFDM transmissions, and therefore uses several frequency sub-carriers. The original Alamouti scheme uses transmissions from two antennas over two time slots: in the first time slot, each antenna transmits a symbol of the data stream, that is, antenna 1 transmits symbol x_1 and antenna 2 transmits symbol x_2. In the second time slot, antenna 1 transmits $-x_2^*$ and antenna 2 transmits x_1^*. For DVB-T2, the Alamouti scheme has been slightly modified, and uses transmissions from two antennas on two subcarriers. Moreover, that is, on the first subcarrier, antenna 1 transmits symbol x_1 and antenna 2 transmits symbol $-x_2^*$, and on the second time slot, antenna 1 transmits x_2 and antenna 2 transmits x_1^*. We observe that this modified scheme allows antenna 1 to transmit symbols of the data stream along the sub-carriers, without introducing any modification (conjugation, multiplication by -1) to them. This has been introduced to guarantee backwards compatibility for receivers that are not DVB-T2 enabled and do not have MIMO decoding logic.

It is expected that in DVB-T2 systems, the channel diversity will result from the use of two orthogonal polarisations. Therefore the important channel parameter is the cross-polarisation discrimination. Initial measurements have shown the median value of the XPR to be around 18 dB for the frequencies of interest, with minimum and maximum values at 9 dB and 25 dB respectively. An additional parameter of interest is the kind of statistics that the channels display (Ricean, known in the DVB-T2 context as F1 channels, Rayleigh, known in the DVB-T2 context at P1 channels, etc.) as well as the fact that not all links will necessarily follow the same statistics.

6.3.2 MIMO in the HF Band

The HF band has been used over the years for a variety of applications, including defense broadcasting, air traffic control, radio location and over the horizon radio surveillance, environmental monitoring and ionospheric research. Indeed HF radio is a cost effective way to establish communications in regions where there is no infrastructure (e.g. over the ocean) or where the infrastructure is limited (e.g. in the third world countries or in high latitude regions).

Research on MIMO communications has concentrated mainly on communications within the UHF band and above, because it is in these frequency regions that the enabling phenomenon of multipath propagation becomes apparent. At HF, radio communications most often rely on line of sight links between the transmitter and the receiver. The objects in the environment are too small (much smaller than the wavelength) to act as reflectors or scatterers.

However, it is well known that HF waves are capable of travelling well beyond line of sight by means of reflections from the ionosphere. These paths may be either single hop or multiple hop reflections from the different ionospheric regions (the E-region and/or the F-region). Additionally, due to the presence of the earth's magnetic field, each incident radio wave splits into two electromagnetic components known as the ordinary (O) and the extraordinary (X) modes. Each magneto-ionic component takes a different path through the ionosphere and arrives at the receiver as elliptically polarised signals with opposite rotational sense. The actual number of multipath components connecting a transmitter to a receiver depends on the geometry of the radio link, the frequency of operation, the time of day, season, electromagnetic activity and sunspot number. Indeed an HF propagation path through the ionosphere is subject to fading because of multipath propagation.

Measurements of MIMO HF systems have shown that the main factor limiting the potential of the use of multiple antenna techniques is the high correlation among the links. Indeed the correlation coefficient drops to 0.5 at an effective antenna separation of approximately 150 m. Other antenna solutions appear to be more suitable then, instead of antenna spacing. Indeed, using polarisation or antenna pattern diversity appears to be the way to apply MIMO techniques in the HF band.

6.3.3   MIMO for Satellite Communications

Satellite systems are characterised by the presence of strong line of sight links between the transmitter and the receiver, and therefore are expected to provide limited advantages in terms of spatial multiplexing gain.

Two scenarios can be considered as MIMO implementations that offer potential advantages in satellite communications.

▶ The first scenario involves the use of both polarisations which can result in an increase of the capacity by a factor of 2. Indeed this scenario is also appealing because satellites are inherently space limited and therefore can only accommodate very compact antenna designs.

▶ The second scenario involves the use of a single terrestrial station and more than one satellite, also known as satellite (or orbital) diversity. This latter scenario has been proposed a means to mitigate the fading induced by rainfall. The main issue in this situation if that the links will have unequal powers: indeed the distances from a single site on earth to various satellites in space can vary significantly, which results in high power imbalance.[2] The result is that the capacity benefit is no longer a multiplicative factor.

Practical measurements have illustrated an interesting trade-off in the use of multiple antenna techniques that attempt to exploit the limited scattering that can only be expected to occur locally around a terrestrial station. In order for the scattering to be significant (and therefore have the benefit of multipath propagation), the terrestrial receiver should be placed in a dense environment such as an urban area. However, in such a case, the received signal to noise ratio suffers a detrimental effect and the resulting capacity is low. In contract, open areas, such as highways, achieve higher capacities despite the lack of scattering, exactly because the received signal power is high.

Beyond the question of suitability of MIMO techniques for satellite communications, there are further practical issues that need to be addressed, the most important of which is the lack of synchronisation among the received signals. The difference in path length is in the order of several kilometers, which in turn makes it necessary for the receiver to have additional circuitry in order to time-align the signals received from different satellites.

6.3.4   Ultrawideband MIMO

Ultrawideband (UWB) systems make use of huge frequency bands from 3.1 GHz to 10.6 GHz in the United States and Asia and at least 6 GHz to 8.5 GHz in Europe. However, the Federal Communications Commission (FCC) in the United States and the European

---

[2] This is analogous to the phenomenon of power imbalance in handheld devices when one antenna is obstructed by the user's body/hair/grip.

Commission in Europe have imposed regulatory masks that limit the permitted power spectral density of the UWB signals to be transmitted. This limitation restricts the achievable data rates and MIMO systems have been proposed as an alternative solution that would additionally facilitate the antenna and amplifier design.

The analysis of UWB systems is different from that of the systems discussed in for example Chapter 2 because the channel statistics are not Gaussian. Indeed the Nakagami, Weibull or lognormal distributions have been found to be more suitable for UWB channels, rather than the Rayleigh or Ricean distributions. Moreover, UWB channels have an exponential power delay profile, and rarely distribute cluster-like behaviour. Finally, it has been shown that in the range of 2.5 times the coherence distance, the antenna correlation follows a pattern of the first kind zero-th order Bessel function of the distance (similar to the uniform angle of arrival case), but a small (approximately 0.4) and constant correlation coefficient is observed for antenna separations larger than that.

An additional difference in the approach used to study UWB channel derives from the power limitations under which they operate. Given the power spectral mask regulations and safety or health concerns, the commonly used assumption of uniform power distribution over the available frequency range is no longer satisfied. Moreover UWB systems operate commonly at very low signal to noise ratios. In these regimes, the behaviour of the capacity expressions introduced in Chapter 2 changes. A basic conclusion in UWB systems is that, if the available SNR is too low, it is better to concentrate the available power on a single antenna or on a single symbol to transmit the information data, rather than to distribute the power across multiple antennas or symbols gain diversity. The diversity gain can only be realised if the available SNR is sufficiently high. Indeed in UWB systems, the prevailing technique appears to be antenna selection that helps compensate for the difference in received signal powers if the shadowing characteristics are different for the links. It is not however necessary to perform antenna selection on a per-tap basis because the number of taps is anyway so large that the small-scale statistics are averaged out even by considering the taps measured at a single antenna.

6.3.5 MIMO for On-body Communications

Many communication systems now are focussing on the personal area network (PAN) which concern the wireless channel around the human body and also on the body itself, known as the body area network (BAN). There are numerous reasons why devices would need to communicate around a body or to a device near to it, such as another BAN or access point for applications including medicine (for monitoring a paitent's medical condition), context aware applications (where the mood of a collection of people or individual needs to be monitored), some possible military applications and also entertainment in virtual reality games. High amounts of data may be necessary to transmit in such circumstances and UWB has indeed been one possibility for this though MIMO also could offer increase in data capacity between two BANs while using less bandwidth. Likewise channel models are required that consider the differing patterns, polarisation and also spatial effects that occur on the body. These would consider antennas that are both attached close to the body onto sensors or inter-woven into textiles in the clothing that the person wears. Results have typically shown there to be a Ricean fading between two bodies in close proximity, though Doppler is highly influenced by the activity and movement of the two bodies. The variable antenna patterns and their positions around the body provide significant MIMO potential.

6.3.6   MIMO for Vehicular Communications

There is now growing interest in vehicle to vehicle communications, most notably the IEEE 802.11p standard for vehicular area networks. There are many reasons for the increased need for communication of data between vehicles on the move as well as from the roadside or other stationary object to a vehicle. Such reasons would include road safety, wireless local area network access on the move to support a greater ubiquitous office environment for people using public transport. With such interests in mind, several measurements have been to date carried out mainly in the 5 GHz band as required by the standard where for the frequency of interest, 5.9 GHz, there can be Doppler shifts therefore up to 710 Hz due to the speed of a vehicle moving at 130 km/h. However, this is not the maximum Doppler shift, since if the speed of two vehicles in opposite directions is concerned, the relative velocity between them therefore increases to 260 km/h and then if a signal bounces off another vehicle moving at 130 km/h, then the maximum Doppler shift is going to be increased three times to as much as 2130 Hz. Clearly the Doppler spread will vary and it will be highly dependent on the environment as well as how many vehicles and other scatters are in the vicinity. This will inevitably change as the distance between vehicles, or between a vehicle and a roadside access point changes.

For vehicle to vehicle MIMO, one setback is the predominantly strong line of sight, which therefore indicates that beamforming is more optimum for delivery of maximum capacity to a vehicle or between vehicles, particularly on rural highways. One important factor, however, is the need to accommodate interference mitigation between two vehicle to vehicle links in such environments which will have highly correlated channel states. Preliminary analysis of vehicle to vehicle results has verified that two links can be made near orthogonal if the antennas are spread over each vehicle. Given that different vehicles will have differently spaced antennas and also different phase weights applied, they will be able to de-correlate two separate vehicle to vehicle links.

6.3.7   MIMO in Small Cellular Environments

One area where MIMO has the potential to be useful is the ever expanding small cells that are being developed, in particular the rapidly growing use of the *femtocell*. A femtocell has initially gained interest with regards to the home environment, where a single home would be considered as a femtocell. The "base station" for the femtocell would be formed by simply connecting a small access point to the ethernet connection already available in the home. Currently femtocells are applied to current 3G mobile though for forthcoming 3G LTE and beyond, as has been clearly seen in this chapter, the requirement to include MIMO will naturally extend to femtocells. It is not only a question of how MIMO will be modelled to the home femtocell, but also to another evolution of femtocells in the outdoor environment, which will operate in areas on the edge of a larger cell or in areas within a larger cell that typically have an intense number of users present, such as an airport concourse, railway station or busy shopping mall. In such areas, several femtocells can be implemented to meet the capacity demands necessary in such areas, which will all need to be MIMO enabled and require potentially suitable ray tracing models that can easily be applied to small areas.

6.4   Concluding Remarks and Future Wireless Systems

This book has aimed to present MIMO communication systems from the perspective of how the multipath in the air interface is exploited to enhance the data throughput of a wireless channel. For this reason the emphasis on channel models has moved away from single independent models that have traditionally been used to determining the interdependence that simultaneous channels have when the transmit antennas are in close proximity as well as the antennas at the receive end. The book hopes to inform the reader about well established MIMO channel modelling techniques as well as how to implement them and allow them to decide on the most appropriate approaches to a specific model they may need to use for a specific application, where they are well aware of the model's advantages as well as its shortcomings.

There are of course other channel models already existent and others that will come into existence for different purposes, which are beyond the scope of this book to include. It is worth noting a number of applications where more advanced channel models will require development, or are under development that this book does not include, some of which are listed below as follows:

▶ **Multi-User MIMO Interference:** In any given wireless system, there is not only one user but many using the same frequency band and so there will be the problem of co-channel interference (at the same frequency) as well as adjacent channel interference (at neighbouring frequencies) (Rahman *et al.* 2007, Webb *et al.* 2004). Therefore in order to test the system's capability with regards to interference mitigation it is necessary to have models that include the simultaneous channels both from the wanted base station and an interfering base station or other user.

▶ **Cooperative MIMO:** In the case of two or more interfering MIMO channel links, they can be seen in one context as interference that needs to be overcome, while on the other hand, the links could work together to enable both users to gain more capacity through doing so. This is known as *cooperation* (Brown, Eggers and Olesen 2007). Therefore the simultaneous channel models used for multi-user MIMO interference are likewise just as valid.

► **MIMO with irregular antenna conditions:** There could be a case for example where the antennas at both the transmit end and receive end are highly directional and as such the scattering conditions between the transmitter and receiver may be fixed and also some scattering objects surrounding the antennas will become irrelevant because the directional antennas will reduce the opportunity for wideband fading.

► **MIMO at high frequencies:** There is a growing interest in using higher frequency bands up to 60 GHz, because of the wide bandwidth that they offer and opportunity to build smaller antennas that will fit into a compact system (Eden *et al.* 2006). However, their disadvantage is that their free space path loss is too large at a short distance and also the signal is vulnerable to absorption by moisture in the air. Therefore such models have to include such factors into the total path loss when used in the context of MIMO.

► **MIMO for radar:** This book has been written with a focus on the wireless communication channel, though MIMO can equally be applied to enhance the returned information from a radar system, particularly with regard to radar imaging applications where the quality can be much improved.

Clearly this list is not exhaustive and it is clear that MIMO channel models will develop as the applications for MIMO develop, which is still opening up several opportunities for current and future research in MIMO channels.

Appendix

Some Useful Definitions

Complex Gaussian random variable. A real valued random vector $X = [x_1, \ldots, x_n]^T$ has a Gaussian distribution if the random variables $x_1, \ldots, x_n$ have a *joint* Gaussian distribution. Assuming that X has a zero mean and covariance matrix $E[XX^T] = R_{xx}$, its probability density function can be written as:

$$f(X) = \frac{1}{\sqrt{(2\pi)^n \det(R_{xx})}} \exp\left[-\frac{1}{2}X^T R_{xx}^{-1} X\right]. \tag{A.1}$$

A complex random variable $Z = \text{Re}(Z) + j\text{Im}(X)$ has a *complex* Gaussian distribution if its real and imaginary parts are jointly Gaussian. Let us define the $2n \times 1$ vector $\mathcal{Z} = \left[\text{Re}(X)^T \quad \text{Im}(X)^T\right]^T$ as the real valued random vector containing the real and imaginary parts of X. The probability density function if $\mathcal{Z} \sim \mathcal{N}(0, R_{zz})$ can be written as in Equation (A.1). When the complex Gaussian random vector Z is additionally circularly symmetric, its distribution can be written in a conveniently compact way as described in the following two items.

Complex circularly symmetric random variable. A complex vectorial random variable Z is said to be circularly symmetric if Z and $e^{j\phi}Z$ have the same distribution for all real values of the phase ϕ.

For a scalar complex random variable, circular symmetry means that its real and imaginary parts are independent and have the same distribution.

A complex circularly symmetric random vector has zero-mean, as $E[Z] = e^{j\phi}E[Z] \Rightarrow (1 - e^{j\phi})E[Z] = 0 \Rightarrow E[Z] = 0$.

Distribution of a complex circularly symmetric Gaussian random variable. Let the vector $Z = [z_1, \ldots, z_n]^T$ be a complex circularly symmetric Gaussian random variable with mean equal to 0 and covariance matrix $E[ZZ^H] = R_{zz}$. Such a circularly

Practical Guide to the MIMO Radio Channel with MATLAB® Examples, First Edition.
Tim Brown, Elisabeth De Carvalho and Persefoni Kyritsi.
© 2012 John Wiley & Sons, Ltd. Published 2012 by John Wiley & Sons, Ltd.

symmetric Gaussian distribution is denoted as $Z \sim \mathcal{CN}(0, R_{zz})$. The probability density function of Z is:

$$f(Z) = \frac{1}{\pi^n \det(R_{zz})} \exp\left[-Z^H R_{zz}^{-1} Z\right].$$ (A.2)

This last expression can be easily extended to a random variable with mean that is non zero. Let us denote as $\mathbf{m}$ the mean of Z and its covariance matrix as $R_{zz} = E[(Z - \mathbf{m})(Z - \mathbf{m})^H]$. If $Z - \mathbf{m}$ is circularly symmetric Gaussian, that is, $Z - \mathbf{m} \sim \mathcal{CN}(0, R_{zz})$, (A.2) can be applied to the random variable $Z - \mathbf{m}$ using a simple change of variable.

MIMO Rayleigh fading model. The MIMO Rayleigh fading model assumes that (a) each subchannel fading h_{ji} has a ZMCCS Gaussian distribution: $h_{ji} \sim \mathcal{CN}(0, \sigma_h^2)$ and b) the subchannel fading processes are independent from each other.

Random or stochastic process. A random or stochastic process is a sequence of random variables $\{x(t), t \in \mathbb{T}\}$ indexed by time and where $\mathbb{T}$ is the set of time indices. It is used to describe the evolution in time of a given process.

For example, a wireless channel can be viewed as a random variable with a certain distribution which depends on the position of the wireless device, the propagation environment, etc.... The value of the channel at a given time is a realisation of the random variable. The evolution in time of the channel distribution is described using a random process. At a given time instant, the wireless device might be in the line of sight of a base station, while, at another time instant, obstacles prevent a line of sight communication: in such a case, the channel distribution at different time instances might change dramatically.

According to the complexity of the channel modelling, the distribution of the random variables might change in time or remain the same. A correlation might also exist between random variables at different time instances. Continuing with the channel process example, a way to model the evolution in time of the channel when the wireless device moves within a local area is as follows. Because the movement is local, the distribution of the channel is assumed to be invariant in time. However, the random variables representing the channel at different time instants are correlated. This correlation factor depends on the mobility of the user. If the user moves slowly compared to the coherence time of the channel, the correlation between two consecutive time instants is high. If it moves fast, the correlation is low.

Two processes $\{x(t), t \in \mathbb{T}\}$ and $\{y(t), t \in \mathbb{T}'\}$ are said to be *independent* if, for all values $t \in \mathbb{T}$, $t' \in \mathbb{T}'$, the random variables $x(t)$ and $y(t')$ are independent.

Stationary stochastic process. A stochastic process is said to be (strictly) stationary if its distribution characteristics are invariant when shifted in time. Consider the stochastic process $\{x(t), t \in \mathbb{T}\}$. $F(x(t_1 + \tau), \ldots, x(t_n + \tau))$ is the joint cumulative distribution function of the stochastic process at time $t_1 + \tau, \ldots, t_n + \tau$. Then, $\{x(t), t \in \mathbb{T}\}$ is said to be stationary if, for all t, τ and n

$$F(x(t_1 + \tau), \ldots, x(t_n + \tau)) = F(x(t_1), \ldots, x(t_n)). \qquad (A.3)$$

Hence, the joint cumulative distribution function is not a function of the time shift τ.

Ergodic stochastic process. Let the stochastic process $\{x(t), t \in \mathbb{T}\}$ and $X(t_1), \ldots, X(t_n)$ be *realisations* of the stochastic process. Consider the time averaging of those realisations: $\frac{1}{n} \sum_{k=1}^{n} X(t_k)$. The stochastic process is said to be ergodic if the time averaging converges to the same limit when n grows to infinity for all realisations $X(t_1), \ldots, X(t_n)$ of the stochastic process. This limit is the ensemble average $E(x)$ and is independent of time:

$$\frac{1}{n} \sum_{k=1}^{n} X(t_k) \stackrel{n \to \infty}{\longrightarrow} E(x). \qquad (A.4)$$

Orthogonal projection matrix. Let us consider the simple geometric example in the figure below: $\mathbf{u}^{\perp}$ is the orthogonal projection of vector $\mathbf{u}$ into the direction of vector $\mathcal{U}$.

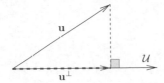

This geometric orthogonal projection can be formalised for a vectorial space of higher dimension with an Hermitian inner product. With this inner product, it is possible to define orthogonality: two vectors $\mathbf{u}$ and $\mathbf{v}$ belonging to the vectorial space are orthogonal is $\mathbf{u}^H \mathbf{v} = 0$. Note that it is possible to define projections that are not orthogonal.

The orthogonal projection of a vectorial space from $\mathbb{C}^M$ (in the figure, vector u) into the column space $\mathcal{U}$ (in the figure, line $\mathcal{U}$) can be represented as a matrix operation using an orthogonal projection matrix.

An orthogonal projection matrix, denoted as P, is a square matrix with the following properties. Let P be of dimension $M \times M$ and let us define $\mathcal{U}$ as the space spanned by the columns of P. Let u be a $M \times 1$ vector belonging to the column space of P. If P is an orthogonal projection matrix, then $Pu = \mathbf{u}$. If $\mathbf{u}$ belongs to the space orthogonal to $\mathcal{U}$, then $P\mathbf{u} = 0$.

A projection matrix is formally defined as a square matrix verifying $P^2 = P$: this relationship means that the column space of P remains intact after the projection. In addition if $P^H = P$, the projection is orthogonal.

P can be written as a function of $\mathcal{U}$. If the columns of $\mathcal{U}$ form an orthonormal basis (i.e. $\mathcal{U}^H \mathcal{U} = I$), an expression for the orthogonal projection matrix into the column space of

$\mathcal{U}$ is $P = \mathcal{U}\mathcal{U}^{H}$. In general, an expression for the orthogonal projection matrix into the column space of $\mathcal{U}$ is $P = \mathcal{U}\left(\mathcal{U}^{H}\mathcal{U}\right)^{-1}\mathcal{U}^{H}$.

Matrix inversion lemma (MIL). Let A, B, C, D be 4 matrices of size $M \times M$, $M \times N$, $N \times N$, $N \times M$ respectively. The matrix inversion lemma states:

$$(A + BCD)^{-1} = A^{-1} - A^{-1}B\left(C^{-1} + DA^{-1}B\right)^{-1}DA^{-1}. \qquad (A.5)$$

A simple application of the matrix inversion lemma is as follows. Suppose A is an identity matrix (or a matrix easy to invert), B is a column vector and D is a row vector. Then, $C^{-1} + DA^{-1}B$ is a scalar. Hence, using the matrix inversion lemma, the computation of the inverse of $A + BCD$ gets simplified.

Bibliography

Alamouti, A. (1998) A simple transmit diversity technique for wireless communication, *IEEE Journal on Selected Areas in Communication* 16(8), (Oct.).

Allen, B, Dohler, M., Okon, E.E., Malik, W.Q., Brown, A.K. and Edwards, D.J. (2007) *Ultra Wideband Antenans and Propagation for Communications, Radar and Imaging*, Chichester: John Wiley & Sons Ltd.

Atanes, P., Arrinda, A., Prieto, G., Angueira, P., Vlez, M.M. and Prieto, P. (2009) MIMO performance of the next generation DVB-T, in *Proc. IEEE 69th Vehicular Technology Conference*, pp. 1–5, Spring.

Bach Andersen, J. (2000a) Antenna arrays in mobile communications: Gain, diversity, and channel capacity, *IEEE Antennas and Propagation Magazine* 42(2): 12–16.

Bach Andersen, J. (2000b) Array gain and capacity for known random channels with multiple element arrays at both ends, *IEEE Journal on Selected Areas in Communications* 18(11) (Nov.): 2172–8,

Bach Andersen, J. (2002) A propagation overview, *Wireless Personal Multimedia Communications Conference* 1: 1–6.

Bach Andersen, J., Nielsen, J.Ø., Pedersen, G.F., Bauch, G, Dietl, G. (2009) Doppler spectrum from moving scatterers in a random environment, *IEEE Transactions on Wireless Communications* 8(6) (June): 3270–7.

Balanis, C.A. (2005) *Antenna Theory, Analysis and Design*, 3rd edn, Chichester: John Wiley & Sons Ltd.

Brown, T.W.C., Eggers, P.C.F. and Olesen, K. (2007) Simultaneous 5 GHz co-channel multiple-input multiple output links at microcellular boundaries: interference or cooperation?, *IET Microwaves, Antennas and Propagation* 1(6): 1152–9.

Brown, T.W.C. and Saunders, S.R. (2007) The intelligent quadrifilar helix: a compact antenna for IEEE 802.11n, *European Conference of Antennas and Propagation (EuCAP)*, Nov.

Brown, T.W.C., Saunders, S.R. and Evans, B.G. (2005) Analysis of mobile terminal diversity antennas, *IEE Proceedings on Microwaves, Antennas and Propagation* 152(1): 1–6.

Brown, T.W.C., Saunders, S.R., Stavrou, S. and Fiacco, M. (2007) Characterization of polarization diversity at the mobile terminal, *IEEE Transactions on Vehicular Technology* 56(5/1): 2440–7.

Chen, Z.N. *et al.* (2007) *Antennas for Portable Devices*, Chichester: John Wiley & Sons Ltd.

Chiau, C.C., Chen, X. and Parini, C.G. (2005) A compact four-element diversity-antenna array for PDA terminals in a MIMO system, *Microwave and Optical Technology Letters*, Wiley Interscience, 44(5) (Jan.): 408–12.

Chizhik, D., Farrokhi, F.R., Ling, J., Lozano, A. (2000) Effect of antenna separation on the capacity of BLAST in correlated channels, *IEEE Communication Letters* 4(11) (Nov.): 334–6.

Practical Guide to the MIMO Radio Channel with MATLAB® Examples, First Edition.
Tim Brown, Elisabeth De Carvalho and Persefoni Kyritsi.
© 2012 John Wiley & Sons, Ltd. Published 2012 by John Wiley & Sons, Ltd.

Chizhik, D., Foschini, G.J., Gans, M.J. and Valenzuela, R.A. (2002) Keyholes, correlations, and capacities of multielement transmit and receive antennas, *IEEE Transactions on Wireless Communications* 1(2) (Apr.): 361–8.

Chizhik, D., Foschini, G.J. and Valenzuela, R.A. (2000) Capacities of multi-element transmit and receive antennas: Correlations and keyholes, *Electronics Letters* 36(13) (June): 1100.

Correia, L.M. *et al.* (2001) *Wireless Flexible Personalised Communications, COST259, European Cooperation in Mobile Radio Research*, Chichester: John Wiley & Sons Ltd.

Correia, L.M. *et al.* (2006) *Mobile Broadband Multimedia Networks*, COST 273 Report, Elsevier Academic Press.

Degli-Esposti, V. (2001) A diffuse scattering model for urban propagation prediction, *IEEE Transactions on Antennas and Propagation* 49(7), July.

Dunman, T.M. and Grayheb, A. (2007) *Coding for MIMO Communication Systems* Chichester: John Wiley & Sons Ltd.

Eden, D., Lockton, A.S. and Chan, A.H.Y. (2006) Preliminary characterisation of the propagation channel at 60GHz for static and mobile links, *European Conference on Antennas and Propagation*, pp. 1–6.

Eggers, P.C.F., Brown, T.W.C., Olesen, K. and Pedersen, G.F. (2007) Assessment of capacity support and scattering in experimental high speed vehicle to vehicle MIMO links, *IEEE Vehicular Technology Conference*, 466–70.

Ertel, R.B. and Reed, J.H. (1999) Angle and time of arrival statistics for circular and elliptical scattering models, *IEEE Journal on Selected Areas in Communications* 17(11) (Nov.): 1829–40.

Farrokhi, F.R., Foschini, G.J., Lozano, A. and Valenzuela, R.A. (2001) Link-optimal space-time processing with multiple transmit and receive antennas, *IEEE Communication Letters* 5(3) (Mar.): 85–7.

Fleury, B.H. (1996) An uncertainty relation for WSS processes and its application to WSSUS systems, *IEEE Transactions on Communications* 44(12): 1632–4.

Foschini, G.J. (1996) Layered space-time architecture for wireless communication in a fading environment when using multi-element antennas, *Bell Laboratories Technical Journal*, 4159, Oct.

Foschini, G.J. and Gans, M.J. (1998) On limits of wireless communications in a fading environment when using multiple antennas, *Wireless Personal Communications* 6(3): 311–35, June.

Foschini, G.J., Golden, G.D., Valenzuela, R.A. and Wolniansky, P.W. (1999) Simplified processing for high spectral efficiency wireless communication employing multi-element arrays, *IEEE Journal on Selected Areas in Communications* 17(11) (Nov.): 1841–52.

Gao, Y., Chen, X. and Parini, C.G. (2007) Channel capacity of dual-element modified PIFA array on small mobile terminal, *Electronics Letters* 43(20): 1060–2.

Gesbert, D., Bolcskei, H., Gore, D.A. and Paulraj, A.J. (2002) Outdoor MIMO wireless channels: models and performance prediction, *IEEE Transactions on Communications* 50(12) (Dec.): 1926–34.

Golden, G.D., Foschini, G.J., Valenzuela, R.A. and Wolniansky, P.W. (1999) Detection algorithm and initial laboratory results using V-BLAST spacetime communication architecture, *Electronics Letters* 35 (Jan.): 1416.

Goldsmith, A. (1997) The capacity of downlink fading channels with variable rate and power, *IEEE Transactions on Vehicular Technology* 46(3) (Aug.): 569–80.

Goldsmith, A. and Chua, S.-G. (1997) Variable-rate variable-power MQAM for fading channels, *IEEE Transactions on Communications* 45(10) (Oct.): 1218–30.

Goldsmith, A. and Chua, S.-G. (1998) Adaptive coded modulation for fading channels, *IEEE Transactions on Communications* 46(5): 595–602, May.

Goldsmith, A. and Varaiya, P. (1997) Capacity of fading channels with channel side information, *IEEE Transactions on Information Theory* 43(6) (Nov.): 1986–92.

Gomez-Calero, C., Cuellar, L., Haro, L. de and Martinez, R. (2009) A 22 novel MIMO testbed for DVB-T2 systems, *IEEE International Symposium on Broadband Multimedia Systems and Broadcasting*.

Gunashekar, S.D., Warrington, E.M., Salous, S., *et al.* (2007) An experimental investigation into the feasibility of MIMO techniques within the HF Band, *European Conference on Antennas and Propagation.*

Gunashekar, S.D., Warrington, E.M., Salous, S., *et al.* (2008) Early results of experiments to investigate the feasibility of employing MIMO techniques in the HF band, *Loughborough Antennas and Propagation Conference, 2008.* LAPC, pp. 161–4.

Horvath, P., Karagiannidis, G.K., King, P.R., Stavrou, S. and Frigyes, I. (2007) Investigations in satellite MIMO channel modeling: accent on polarization, *EURASIP Journal on Wireless Communications and Networking* 1.

Huang, Y. and Boyle, K. (2008) *Antennas: From Theory to Practice*, Chichester: John Wiley & Sons Ltd.

Jakes, W.C.(1974) Microwave mobile communications, *IEEE Press.*

Jiang, D. and Delgrossi, L. (2008) IEEE 802.11p: Towards an international standard for wireless access in vehicular environments, *IEEE Vehicular Technology Conference*, 2036–40, May.

Kaiser, T., Feng Zheng, Dimitrov, E. (2009) An overview of ultra-wide-band systems with MIMO, *Proceedings of the IEEE* 97(2) (Feb.): 285–312.

Karedal, K., Tufvesson, F. and Czink, N. *et al.* (2009) Geometry-based stochastic MIMO model for vehicle to vehicle communications, *IEEE Transactions on Wireless Communications* 8(7) (July): 3646–57.

Kermoal, J.P., Schumacher, L., Pedersen, K.I., Mogensen, P.E. and Frederiksen, F. (2002) A stochastic MIMO radio channel model with experimental validation, *IEEE Journal on Selected Areas in Communications* 20(6) (Aug.): 1211–26.

Khan, I. and Hall, P.S. (2009) Experimental evaluation of MIMO Capacity and Correlation for Narrow-band body-centric wireless channels, *IEEE Transactions on Antennas and Propagation* 58(1) (June): 195.

Klemp, O. and Eul, H. (2005) Radiation pattern analysis of antenna systems for MIMO and diversity configurations, *Advances in Radio Science* 3: 157–65.

King, P.R. and Stavrou, S. (2005) Land mobile-satellite MIMO capacity predictions, IEE Electronics Letters 41(13) (June): 749–51.

King, P.R. and Stavrou, S. (2006) Capacity improvement for a land mobile single satellite MIMO system, *IEEE Antennas and Wireless Propagation Letters* 5: 98–100 (Dec.).

King, P.R. and Stavrou, S. (2007) Low elevation wideband land mobile satellite MIMO channel characteristics, *IEEE Transactions on Wireless Communications* 6(7) (July): 2712–20.

Larsson, E. and Stoica, P. (2003) *Space–time Block Coding for Wireless Communications*, Cambridge: Cambridge University Press, 2003.

Ling, J., Chizhik, D. and Valenzuela, R.A. (2001) Predicting multi-element receive and transmit array capacity outdoors with ray tracing, *Proceedings of the IEEE 53rd Vehicular Technology Conference, Spring 2001*, vol. 1, pp. 392–94, May.

Loyka, S. and Tsoulos, G. (2002) Estimating MIMO system performance using the correlation matrix approach, *IEEE Communications Letters* 6(1) (Jan.): 19–21.

Molisch, A.F. (2004) A generic model for MIMO wireless propagation channels in macro- and microcells, *IEEE Transactions on Signal Processing* 52(1) (Jan.): 61–71.

Molisch, A.F. (2005) *Wireless Communications*, Chichester: John Wiley & Sons Ltd.

Ouyang, Y., Love, D.J. and Chappell, W.J. (2009) Body-worn distributed MIMO system, *IEEE Transactions on Vehicular Technology* 58(4) (Apr.): 1752–65.

Ozcelik, M., Czink, N. and Bonek, E. (2005) *IEEE Vehicular Technology* Conference, vol. 1, pp. 156–160, Spring.

Paier, A., Karedal, J., Czink, N. *et al.* (2009) Characterization of vehicle-to-vehicle radio channels from measurements at 5.2 GHz, *Wireless Personal Communications* 50(1) (July): 19–32.

Papoulis, A. (1965) *Probability, Random Variables, and Stochastic Processes*, New York: McGraw-Hill, 9th edn.

Parsons, J.D. (2000) *The Mobile Radio Propagation Channel*, Chichester: John Wiley & Sons Ltd.

Paulraj, A., Nabar, R. and Gore, D.(2003) *Introduction to Space–Time Wireless Communications*, Cambridge: Cambridge University Press.

Proakis, J.D. (2008) *Digital Communications*, Maidenhead: McGraw-Hill, UK, 5th edn.

Rahman, M.I., Carvalho, E. de and Prasad, R. (2007) Impact of MIMO co-channel interference, *IEEE Personal, Indoor and Mobile Radio Communications*, pp. 1–5, 3–7 September.

Roy, R.H., Paulraj, A.J. and Kailath, T. (1990) *Methods and Arrangements for Signal Reception and Parameter Estimation*, US Patent No. 4965732, 23 Oct.

Saleh, A. and Valenzuela, R. (1987) A statistical model for indoor multipath propagation, *IEEE Journal on Selected Areas in Communications* 5(2) (Feb.): 128–37.

Sayeed, A.M. (2002) Deconstructing multiantenna fading channels, *IEEE Transactions on Signal Processing* 50(10): 2563–79, Oct.

Saunders, S.R. and Aragon-Zavala, A. (2007) *Antennas and Propagation for Wireless Communication Systems*, Chichester: John Wiley & Sons Ltd.

Saunders, S.R., Carlaw, S., Giustina, A., Bhat, R.R., Rao, V.S. and Siegberg, R. (2009) *Femtocells: Opportunities and Challenges for Business and Technology*, Chichester: John Wiley & Sons Ltd.

Schumacher, L., Berger, L.T. and Ramiro-Moreno, J. (2002) Recent advances in propagation characterisation and multiple antenna processing in the 3GPP framework, *Proceedings of the URSI Radio Conference*, Aug.

Seshadri, N., Tarokh, V. and Calderbank, A.R. (1997) Space-time codes for wireless communication: code construction, *IEEE Vehicular Technology Conference* 2: 637–41, May.

Shannon, C.E. (1948) A mathematical theory of communication, *Bell System Technical Journal* 27: 379–423, 623–56, July and Oct.

Shiu, D.-S., Foschini, G.J., Gans, M.J. and Kahn, J.M. (2000) Fading correlation and its effect on the capacity of multielement antenna systems, *IEEE Transactions on Communications* 48(3), Mar.

Steinbauer, M., Molisch, A.F. and Bonek, E. (2001) The double-directional radio channel, *IEEE Antennas and Propagation Magazine* 43(4): 51–3, Aug.

Strang, G. (2009) *Introduction to Linear Algebra*, Wellesley-Cambridge Press.

Sulonen, K., Vainikainen, P. and Kallilola, K. (2002) The effect of angular power distribution in different environments and the angular resolution of radiation pattern measurement on antenna performance, *COST273 Temporary Document*, TD(02)028, Jan.

Suvikunnas, P., Villanen, J., Sulonen, K., Icheln, C., Ollikainen, J. and Vainkainen, P. (2006) Evaluation of the performance of multiantenna terminals using a new approach, *IEEE Transactions on Instrumentation and Measurement* 55(5): 1804–13, Oct.

Taga, T. (1990) Analysis for mean effective gain of mobile antennas in land mobileradio environments, *IEEE Transactions on Vehicular Technology* 39(2): 117–31, May.

Tarrokh, V., Jafarkhani, H. and Calderbank, A.R. (1999) Space–time block codes from orthogonal designs, *IEEE Transactions on Information Theory* 48(5) (May): 1456–67.

Tarrokh, V., Seshadri, N. and Calderbank, A.R. (1998) Space–time codes for high data rate wireless communications: performance criterion and code construction, *IEEE Transactions on Information Theory* 44(2) (Feb.): 744–65.

Telatar, E. (1999) Capacity of multi-antenna Gaussian channels, *European Transactions on Telecommunications* 10(6): 585–96.

Telatar, T.E. and Tse, D.N.C. (2000) Capacity and mutual information of wideband multipath fading channels, *IEEE Transactions on Information Theory* 46(4) (July): 1384–1400.

Tran, V.P. and Sibille, A. (2006) Spatial multiplexing in UWB MIMO communications, *Electronics Letters* 42(16): 931–2.

Tse, D., Viswanath, P. and Zheng, L. (2004) Diversity-multiplexing tradeoff in multiple access channels , *IEEE Transactions on Information Theory* 50(9) (Sep.): 1859–74.

Vaughan, R.G. and Bach Andersen, J. (1987) Antenna diversity in mobile communications, *IEEE Transactions on Vehicular Technology* 36(4) (Apr.): 149–72.

Vaughan, R.G. and Bach Andersen, J. (2002) *Channels, Propagation and Antennas for Mobile Communications,* IEE Publishing.

Waldschmidt, C. and Wiesbeck, W. (2004) Compact wide-band multimode antennas for MIMO and diversity, *IEEE Transactions on Antennas and Propagation* 52(8) (Aug.): 1963 9.

Wang, C.-X., Cheng, X. and Laurenson, D.I. (2009) Vehicle-to-vehicle channel modeling and measurements: recent advances and future challenges, *IEEE Communications Magazine* 47(11) (Nov.): 96–103.

Webb, M., Beach, M. and Nix, A. (2004) Capacity limits of MIMO channels with co-channel interference, *IEEE Vehicular Technology Conference*, May, vol. 2, pp. 703–7.

Weichselberger, W., Herdin, M., Özcelik, M. and Bonek, E. (2006) A stochastic MIMO channel model with joint correlation of both link ends, *IEEE Transactions on Wireless Communications* 5(1), Jan.

Winters, J.H. (1987) On the capacity of radio communication systems with diversity in a Rayleigh fading environment, *IEEE Journal on Selected Areas in Communications* 5(5) (June): 871–8.

Wolniansky, P.W., Foschini, G.J., Golden, G.D. and Vlenzuela, R.A. (1998) V-BLAST: an architecture for realizing very high data rates over the ruch-scattering wireless channel, *Proceedings of the URSI International Symposium on Signals, Systems and Electronics Conference*, pp. 295–300.

Xu, H., Chizhik, D. , Huang, H. and Valenzuela, R. (2004) A generalized space–time multiple-input multiple-output (MIMO) channel model, *IEEE Transactions on Wireless Communications* 3(3) (May): 966–75.

Zhiyi, H. and.Wei, C (2008) A T-intersection street model of MIMO vehicle-to-vehicle fading channel, *International Conference on Communications, Circuits and Systems, 2008*, May, pp. 75–

Websites

http://standards.ieee.org/getieee802/ – IEEE 802.16m-08/004r5, 'IEEE 802.16m Evaluation Methodology Document (EMD)', *IEEE 802.16 Broadband Wireless Access Working Group*, Jan. 2009.

http://www.3gpp.org – 3rd Generation Partnership Project, Technical Specification Group Radio Access Network, 'Spatial channel model for Multiple Input Multiple Output (MIMO)Simulations', *3GPP TR 25.996*, v8.0.0, 2008.1

Index

Practical Guide to the MIMO Radio Channel with MATLAB® Examples, First Edition.
Tim Brown, Elisabeth De Carvalho and Persefoni Kyritsi.
© 2012 John Wiley & Sons, Ltd. Published 2012 by John Wiley & Sons, Ltd.